Official SQA Past Papers WITH ANSWERS

Advanced Higher
Physics

2010–2014

Hodder Gibson Study Skills Advice – General — page 3
Hodder Gibson Study Skills Advice –
  Advanced Higher Physics — page 5
2010 EXAM — page 9
2011 EXAM — page 33
2012 EXAM — page 53
2013 EXAM — page 77
2014 EXAM — page 99
ANSWER SECTION — page 121

HODDER
GIBSON
AN HACHETTE UK COMPANY

Hodder Gibson is grateful to the copyright holders, as credited on the final page of the Question Section, for permission to use their material. Every effort has been made to trace the copyright holders and to obtain their permission for the use of copyright material. Hodder Gibson will be happy to receive information allowing us to rectify any error or omission in future editions.

Hachette UK's policy is to use papers that are natural, renewable and recyclable products and made from wood grown in sustainable forests. The logging and manufacturing processes are expected to conform to the environmental regulations of the country of origin.

Orders: please contact Bookpoint Ltd, 130 Park Drive, Abingdon, Oxon OX14 4SE. Telephone: (44) 01235 827720. Fax: (44) 01235 400454.

Lines are open 9.00–5.00, Monday to Saturday, with a 24-hour message answering service. Visit our website at www.hoddereducation.co.uk. Hodder Gibson can be contacted direct on: Tel: 0141 848 1609; Fax: 0141 889 6315; email: hoddergibson@hodder.co.uk

This collection first published in 2014 by

Hodder Gibson, an imprint of Hodder Education,

An Hachette UK Company

2a Christie Street

Paisley PA1 1NB

{BrightRED   Hodder Gibson is grateful to Bright Red Publishing Ltd for collaborative work in preparation of this book and all SQA Past Paper, National 5 and Higher for CfE Model Paper titles 2014.

Typeset by PDQ Digital Media Solutions Ltd, Bungay, Suffolk NR35 1BY

Printed in the UK

A catalogue record for this title is available from the British Library

ISBN 978-1-4718-3675-6

3 2 1

2015 2014

Introduction

Study Skills – what you need to know to pass exams!

Pause for thought

Many students might skip quickly through a page like this. After all, we all know how to revise. Do you really though?

Think about this:

"IF YOU ALWAYS DO WHAT YOU ALWAYS DO, YOU WILL ALWAYS GET WHAT YOU HAVE ALWAYS GOT."

Do you like the grades you get? Do you want to do better? If you get full marks in your assessment, then that's great! Change nothing! This section is just to help you get that little bit better than you already are.

There are two main parts to the advice on offer here. The first part highlights fairly obvious things but which are also very important. The second part makes suggestions about revision that you might not have thought about but which WILL help you.

Part 1

DOH! It's so obvious but …

Start revising in good time

Don't leave it until the last minute – this will make you panic.

Make a revision timetable that sets out work time AND play time.

Sleep and eat!

Obvious really, and very helpful. Avoid arguments or stressful things too – even games that wind you up. You need to be fit, awake and focused!

Know your place!

Make sure you know exactly **WHEN and WHERE** your exams are.

Know your enemy!

Make sure you know what to expect in the exam.

How is the paper structured?

How much time is there for each question?

What types of question are involved?

Which topics seem to come up time and time again?

Which topics are your strongest and which are your weakest?

Are all topics compulsory or are there choices?

Learn by DOING!

There is no substitute for past papers and practice papers – they are simply essential! Tackling this collection of papers and answers is exactly the right thing to be doing as your exams approach.

Part 2

People learn in different ways. Some like low light, some bright. Some like early morning, some like evening / night. Some prefer warm, some prefer cold. But everyone uses their BRAIN and the brain works when it is active. Passive learning – sitting gazing at notes – is the most INEFFICIENT way to learn anything. Below you will find tips and ideas for making your revision more effective and maybe even more enjoyable. What follows gets your brain active, and active learning works!

Activity 1 – Stop and review

Step 1

When you have done no more than 5 minutes of revision reading STOP!

Step 2

Write a heading in your own words which sums up the topic you have been revising.

Step 3

Write a summary of what you have revised in no more than two sentences. Don't fool yourself by saying, "I know it, but I cannot put it into words". That just means you don't know it well enough. If you cannot write your summary, revise that section again, knowing that you must write a summary at the end of it. Many of you will have notebooks full of blue/black ink writing. Many of the pages will not be especially attractive or memorable so try to liven them up a bit with colour as you are reviewing and rewriting. **This is a great memory aid, and memory is the most important thing.**

Activity 2 — Use technology!

Why should everything be written down? Have you thought about "mental" maps, diagrams, cartoons and colour to help you learn? And rather than write down notes, why not record your revision material?

What about having a text message revision session with friends? Keep in touch with them to find out how and what they are revising and share ideas and questions.

Why not make a video diary where you tell the camera what you are doing, what you think you have learned and what you still have to do? No one has to see or hear it, but the process of having to organise your thoughts in a formal way to explain something is a very important learning practice.

Be sure to make use of electronic files. You could begin to summarise your class notes. Your typing might be slow, but it will get faster and the typed notes will be easier to read than the scribbles in your class notes. Try to add different fonts and colours to make your work stand out. You can easily Google relevant pictures, cartoons and diagrams which you can copy and paste to make your work more attractive and **MEMORABLE**.

Activity 3 – This is it. Do this and you will know lots!

Step 1

In this task you must be very honest with yourself! Find the SQA syllabus for your subject (www.sqa.org.uk). Look at how it is broken down into main topics called MANDATORY knowledge. That means stuff you MUST know.

Step 2

BEFORE you do ANY revision on this topic, write a list of everything that you already know about the subject. It might be quite a long list but you only need to write it once. It shows you all the information that is already in your long-term memory so you know what parts you do not need to revise!

Step 3

Pick a chapter or section from your book or revision notes. Choose a fairly large section or a whole chapter to get the most out of this activity.

With a buddy, use Skype, Facetime, Twitter or any other communication you have, to play the game "If this is the answer, what is the question?". For example, if you are revising Geography and the answer you provide is "meander", your buddy would have to make up a question like "What is the word that describes a feature of a river where it flows slowly and bends often from side to side?".

Make up 10 "answers" based on the content of the chapter or section you are using. Give this to your buddy to solve while you solve theirs.

Step 4

Construct a wordsearch of at least 10 X 10 squares. You can make it as big as you like but keep it realistic. Work together with a group of friends. Many apps allow you to make wordsearch puzzles online. The words and phrases can go in any direction and phrases can be split. Your puzzle must only contain facts linked to the topic you are revising. Your task is to find 10 bits of information to hide in your puzzle, but you must not repeat information that you used in Step 3. DO NOT show where the words are. Fill up empty squares with random letters. Remember to keep a note of where your answers are hidden but do not show your friends. When you have a complete puzzle, exchange it with a friend to solve each other's puzzle.

Step 5

Now make up 10 questions (not "answers" this time) based on the same chapter used in the previous two tasks. Again, you must find NEW information that you have not yet used. Now it's getting hard to find that new information! Again, give your questions to a friend to answer.

Step 6

As you have been doing the puzzles, your brain has been actively searching for new information. Now write a NEW LIST that contains only the new information you have discovered when doing the puzzles. Your new list is the one to look at repeatedly for short bursts over the next few days. Try to remember more and more of it without looking at it. After a few days, you should be able to add words from your second list to your first list as you increase the information in your long-term memory.

FINALLY! Be inspired...

Make a list of different revision ideas and beside each one write **THINGS I HAVE** tried, **THINGS I WILL** try and **THINGS I MIGHT** try. Don't be scared of trying something new.

And remember – "FAIL TO PREPARE AND PREPARE TO FAIL!"

Advanced Higher Physics

The course

The course comprises two whole and two half units:

- Mechanics (40 Hours)
- Electrical Phenomena (40 hours)
- Wave Phenomena (20 hours)
- Investigation (20 hours)

Course assessment

In order to gain an award in the course, you must pass a NAB in each topic. The NABs are internally assessed by the teacher/lecturer and the evidence is externally verified.

Investigation report

To obtain a pass in the NAB for this unit, you must keep a record or diary of your planning, measurements taken, uncertainties, uncertainty calculations and any notes that might be useful when writing up your report.

This record could be externally verified by SQA and should be submitted to your teacher once the investigation has been completed.

The final Investigation report will be sent to SQA and externally marked out of 25 marks. The final grade awarded for the course will depend on the total marks obtained by the candidate (out of 125) for the examination and the Investigation report.

Advanced Higher Physics is an excellent qualification to have and, although it is a step above Higher, is proof that you have the ability to do well. It just takes hard work and application – the danger is to have too many distractions in your sixth year. Be focused and ensure you are on top of the work immediately. Try to choose an investigation that suits and **stick to deadlines**! Above all, enjoy the course and make sure you stay on the right track.

The examination

You will sit an **examination** lasting 2 hours 30 minutes with an allocation of 100 marks.

Mark allocation

- **3 or 4 mark questions** will generally involve more than one step or several points of coverage.
- **2 mark questions** will involve just one use of an equation or a couple of descriptive points.

Use of the Data Sheet

Try to clearly show where you have substituted a value from the data sheet. For example, **do not** leave μ_0 in an equation. You must show the substitution $4\pi \times 10^{-7}$ in your equation.

When rounding, **do not** round the given values. For example, the mass of a proton = 1.673×10^{-27} kg, **NOT** 1.67×10^{-27} kg.

Use of the Data Booklet

Although many of the required equations are given, it is better to learn the basic equations to save time in the examination.

Remember that the Advanced Higher information is contained on pages 6 to 8. Page 8 gives the moments of inertia of different shapes, plus mathematical relationships.

Solving equations

Do not rearrange equations in algebraic form. Select the appropriate equation, substitute the given values then rearrange the equation to obtain the required unknown. This minimises the risk of wrong substitution.

"Show" questions

Generally, **all steps** for these must be shown, even though they might seem obvious. **Do not assume that substitutions are obvious to the marker.**

All equations used must be stated separately and then clearly substituted if required. Many candidates will look at the end product and somehow end up with the required answer. The marker will be checking that the path to the solution is clear. It is good practice to state why certain equations are used and explain the Physics behind them. For example, problems involving gravitational force supplying the centripetal force should start with this statement:

Calculus – equations of motion

$s = f(t)$

Be clear that differentiating once gives the velocity, differentiating twice gives the acceleration.

$a = f(t)$

Integrating once gives the velocity, integrating twice gives the displacement. Remember to take into account the constant of integration each time by considering the limits.

Definitions

Know and understand definitions given in the course. Definitions often come from the interpretation of an equation.

Diagrams

Use a ruler and appropriate labels. Angles will be important in certain diagrams. Too many candidates attempt to draw ray diagrams freehand and lose marks for untidiness.

Graphs

Read the question and ensure you know what is being asked. Label your graph correctly and do not forget to label the origin.

"Explain" and "describe" questions

These tend to be done poorly. Ensure all points are covered and read over again in order to check that there are no mistakes. Try to be clear and to the point, highlighting the relevant points of Physics.

Uncertainties

In this area, you must understand the following:

- Systematic, scale reading (analogue and digital) and random uncertainties.
- Percentage/Fractional uncertainties.
- Combinations – Pythagorean relationship.
- Absolute uncertainty in final answer (give to one significant figure).

Handling data

An unfamiliar equation or measurements from an experiment might be given. Don't panic, just read the instructions. Relationships between the quantities can be found graphically or algebraically.

Improvements to experimental procedure

Look at the percentage uncertainties in the measured quantities and decide which is most significant. Suggest how the size of this uncertainty could be reduced – do not suggest using better apparatus! It might be better to repeat readings, so that random uncertainty is reduced, or increase distances to reduce the percentage uncertainty in scale reading. It really depends on the experiment, so you will need to use your judgement.

Common points to look out for

Below are common points missed by candidates in each unit. Obviously, all points in all units cannot be covered, but hopefully the following can give you a start in what to look out for. It is important to read the Principal Assessor's report that can be downloaded from SQA's website www.sqa.org.uk. This will give more detail on the performance of candidates in previous exams.

Unit 1

Equations of motion

Ensure you are comfortable with the derivations of the equations of motion using calculus. Remember to account for the constants of integration, if appropriate.

Circular motion

There will often be a question on **central** or **centripetal force**. Remember that this is the force that acts on an object causing the object to follow a circular path.

There is no outward (centrifugal) force acting on that object and it would be good advice **not to mention centrifugal force** in your description.

It is worth noting that a fun fair ride might give the impression of feeling an outward force acting on a person, but this is an unreal force – they just think the force is outward. In fact, it is the inward force from the seat that enables the person to follow the circular path.

The same will apply with a "loop the loop" ride where, at the top of the loop, the track and weight of the car supply the required centripetal force.

Circular motion and planetary motion

The key to calculating the period, T, of motion of a planet or satellite lies with gravitational force supplying the central force. For example:

$GMm / r^2 = mr\omega^2$ then use $= 2\pi / T$ to find T.

Kinetic energy of a satellite

Many candidates look for the normal equation for kinetic energy, but there is an alternative. They should be able to derive and use:

$$E_k = 1/2\ mv^2 = GMm / 2r$$

Simple harmonic motion – know the definition of SHM

Be able to derive the velocity (differentiate once) and acceleration (differentiate twice), given the expression for displacement. For example:

$$y = A\sin\omega t \quad \text{or} \quad x = A\cos\omega t$$

From v and a, the expressions for the maximum velocity and maximum acceleration can be found. (These occur when the maximum value of cos or sin = 1.)

Escape velocity derivation

The starting point is the realisation when the object has escaped the pull of gravity **then $E_T = 0$**.

For example: $E_T = E_K + E_P = 1/2mv^2 - GMm/r = 0$

Rearrange the equation to obtain the **expression for v**.

(Note $E_P = -GMm/r$)

Bohr model of the atom

Many candidates omit that the angular momentum of the electrons is quantised.

Unit 2

Electric field strength

Parallel Plates   **Point Charge**

(Uniform field between plates)

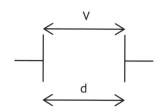   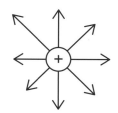

Candidates often mix up these formulae for electric field strength. They are quite different!

"Distance of closest approach' questions"

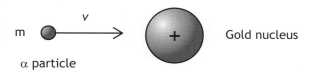

The number of protons in a gold nucleus (or any element) can be found from the periodic table in the data booklet. (Gold has 79 protons.)

Remember, an α particle is a helium nucleus with an charge of $+(2 \times 1.6 \times 10^{-19})$ C.

To solve the problem, use:

$$\tfrac{1}{2} mv^2 = Q_1Q_2/4\pi\varepsilon_0 r$$

To solve for r:

$Q_1 = 3.2 \times 10^{-19}$ C $Q_2 = 79 \times 1.6 \times 10^{-19}$ C

The mass of the α particle is given in the data sheet.

Charged particles in a magnetic field

A circular orbit will be produced if the charged particle cuts the field at 90°.

The central force is produced by the magnetic force acting on the particle.

$$mv^2 / r = Bqv$$

If the charge enters the field at an angle of **less than 90°**, then the resultant path will be **helical**.

An example of this would be charged particles being deflected by the Earth's magnetic field to the North or South poles, producing the borealis.

Back Emf in an Inductor (coil)

$$E = -L \, dI/dt$$

Remember that the back Emf is produced by a changing current, which in turn produces a **changing magnetic field** throughout the coil. Many candidates omit this in their explanation.

Be aware that if asked to calculate the inductance L (or the rate of change of current) and a negative answer is obtained, then **there has been an error. Invariably, the fact that the induced Emf is negative has not been taken into account**.

For example:

E = 9V   dI/dt = 5 As⁻¹

E = − L dI/dt

9 = − L x 5

L = − 1.5 H

This is incorrect!

E = − 9V   dI/dt = 5 As⁻¹

E = − L dI/dt

− 9 = − L × 5

L = 1.5 H

This is correct!

Unit 3

The Doppler effect

Candidates should be able to derive all the equations, whether it be:

(a) the source travelling towards the observer, or

(b) the observer travelling towards the source.

The derivation(s) should be accompanied by an explanation and not just a collection of equations. Diagrams can be useful.

Standing or stationary waves

Remember, these are produced when a reflected wave interferes with the incident wave causing maxima (antinodes) and minima (nodes).

Intensity of a wave

Many candidates are not confident in the use of the relationship between the intensity of a wave and its amplitude.

$$I \propto A^2$$

This can be expressed as

$$I_1 / A_1^2 = I_2 / A_2^2$$

Example $A_1 = 0.04$ m and the intensity increases by times 5.

Calculate the new amplitude.

$$I_1 / 0.04^2 = 5I_1 / A_2^2$$

$$A_2^2 = 5 \times 0.04^2 = 0.09 \text{ m}$$

Note: the amplitude has increased by a factor of $\sqrt{5}$.

Remember, more information on questions and the Investigation report are in the Principal Assessor's External Report, which can be downloaded from www.sqa.org.uk/sqa/40814.html.

Good luck!

Remember that the rewards for passing Advanced Higher Physics are well worth it! Your pass will help you get the future you want for yourself. In the exam, be confident in your own ability. If you're not sure how to answer a question, trust your instincts and just give it a go anyway. Keep calm and don't panic! GOOD LUCK!

[BLANK PAGE]

X069/701

NATIONAL
QUALIFICATIONS
2010

FRIDAY, 28 MAY
1.00 PM – 3.30 PM

PHYSICS
ADVANCED HIGHER

Reference may be made to the Physics Data Booklet.

Answer **all** questions.

Any necessary data may be found in the Data Sheet on page two.

Care should be taken to give an appropriate number of significant figures in the final answers to calculations.

Square-ruled paper (if used) should be placed inside the front cover of the answer book for return to the Scottish Qualifications Authority.

DATA SHEET
COMMON PHYSICAL QUANTITIES

Quantity	Symbol	Value	Quantity	Symbol	Value
Gravitational acceleration on Earth	g	$9 \cdot 8 \text{ m s}^{-2}$	Mass of electron	m_e	$9 \cdot 11 \times 10^{-31} \text{ kg}$
Radius of Earth	R_E	$6 \cdot 4 \times 10^6 \text{ m}$	Charge on electron	e	$-1 \cdot 60 \times 10^{-19} \text{ C}$
Mass of Earth	M_E	$6 \cdot 0 \times 10^{24} \text{ kg}$	Mass of neutron	m_n	$1 \cdot 675 \times 10^{-27} \text{ kg}$
Mass of Moon	M_M	$7 \cdot 3 \times 10^{22} \text{ kg}$	Mass of proton	m_p	$1 \cdot 673 \times 10^{-27} \text{ kg}$
Radius of Moon	R_M	$1 \cdot 7 \times 10^6 \text{ m}$	Mass of alpha particle	m_α	$6 \cdot 645 \times 10^{-27} \text{ kg}$
Mean Radius of Moon Orbit		$3 \cdot 84 \times 10^8 \text{ m}$	Charge on alpha particle		$3 \cdot 20 \times 10^{-19} \text{ C}$
Universal constant of gravitation	G	$6 \cdot 67 \times 10^{-11} \text{ m}^3 \text{ kg}^{-1} \text{ s}^{-2}$	Planck's constant	h	$6 \cdot 63 \times 10^{-34} \text{ J s}$
Speed of light in vacuum	c	$3 \cdot 0 \times 10^8 \text{ m s}^{-1}$	Permittivity of free space	ε_0	$8 \cdot 85 \times 10^{-12} \text{ F m}^{-1}$
Speed of sound in air	v	$3 \cdot 4 \times 10^2 \text{ m s}^{-1}$	Permeability of free space	μ_0	$4\pi \times 10^{-7} \text{ H m}^{-1}$

REFRACTIVE INDICES

The refractive indices refer to sodium light of wavelength 589 nm and to substances at a temperature of 273 K.

Substance	Refractive index	Substance	Refractive index
Diamond	2·42	Glycerol	1·47
Glass	1·51	Water	1·33
Ice	1·31	Air	1·00
Perspex	1·49	Magnesium Fluoride	1·38

SPECTRAL LINES

Element	Wavelength/nm	Colour	Element	Wavelength/nm	Colour
Hydrogen	656	Red	Cadmium	644	Red
	486	Blue-green		509	Green
	434	Blue-violet		480	Blue
	410	Violet		Lasers	
	397	Ultraviolet	Element	Wavelength/nm	Colour
	389	Ultraviolet	Carbon dioxide	9550 ⎱ 10590 ⎰	Infrared
Sodium	589	Yellow	Helium-neon	633	Red

PROPERTIES OF SELECTED MATERIALS

Substance	Density/ kg m^{-3}	Melting Point/ K	Boiling Point/ K	Specific Heat Capacity/ J kg^{-1} K^{-1}	Specific Latent Heat of Fusion/ J kg^{-1}	Specific Latent Heat of Vaporisation/ J kg^{-1}
Aluminium	$2 \cdot 70 \times 10^3$	933	2623	$9 \cdot 02 \times 10^2$	$3 \cdot 95 \times 10^5$	....
Copper	$8 \cdot 96 \times 10^3$	1357	2853	$3 \cdot 86 \times 10^2$	$2 \cdot 05 \times 10^5$	....
Glass	$2 \cdot 60 \times 10^3$	1400	....	$6 \cdot 70 \times 10^2$	....	....
Ice	$9 \cdot 20 \times 10^2$	273	....	$2 \cdot 10 \times 10^3$	$3 \cdot 34 \times 10^5$	....
Glycerol	$1 \cdot 26 \times 10^3$	291	563	$2 \cdot 43 \times 10^3$	$1 \cdot 81 \times 10^5$	$8 \cdot 30 \times 10^5$
Methanol	$7 \cdot 91 \times 10^2$	175	338	$2 \cdot 52 \times 10^3$	$9 \cdot 9 \times 10^4$	$1 \cdot 12 \times 10^6$
Sea Water	$1 \cdot 02 \times 10^3$	264	377	$3 \cdot 93 \times 10^3$	....	....
Water	$1 \cdot 00 \times 10^3$	273	373	$4 \cdot 19 \times 10^3$	$3 \cdot 34 \times 10^5$	$2 \cdot 26 \times 10^6$
Air	1·29	....	....	....	....	....
Hydrogen	$9 \cdot 0 \times 10^{-2}$	14	20	$1 \cdot 43 \times 10^4$	....	$4 \cdot 50 \times 10^5$
Nitrogen	1·25	63	77	$1 \cdot 04 \times 10^3$	....	$2 \cdot 00 \times 10^5$
Oxygen	1·43	55	90	$9 \cdot 18 \times 10^2$	....	$2 \cdot 40 \times 10^5$

The gas densities refer to a temperature of 273 K and a pressure of $1 \cdot 01 \times 10^5$ Pa.

[BLANK PAGE]

Marks

1. A turntable, radius r, rotates with a constant angular velocity ω about an axis of rotation. Point X on the circumference of the turntable is moving with a tangential speed v, as shown in Figure 1A.

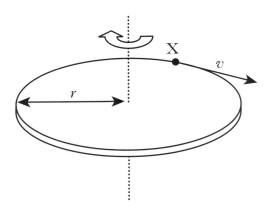

Figure 1A

(a) Derive the relationship

$$v = r\,w.$$

2

(b) Data recorded for the turntable is shown below.

Angle of rotation	$(3\cdot1 \pm 0\cdot1)$ rad
Time taken for angle of rotation	$(4\cdot5 \pm 0\cdot1)$ s
Radius of disk	$(0\cdot148 \pm 0\cdot001)$ m

   (i) Calculate the tangential speed v.   2

   (ii) Calculate the percentage uncertainty in this value of v.   2

   (iii) As the disk rotates, v remains constant.

       (A) Explain why point X is accelerating.   1

       (B) State the direction of this acceleration.   1

   (8)

Marks

2. A motorised model plane is attached to a light string anchored to a ceiling.

   The plane follows a circular path of radius 0·35 m as shown in Figure 2A.

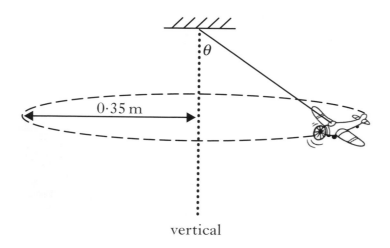

Figure 2A

The plane has a mass of 0·20 kg and moves with a constant angular velocity of 6·0 rad s^{-1}.

(a) Calculate the central force acting on the plane. 2

(b) Calculate angle θ of the string to the vertical. 2

(c) What effect would a decrease in the plane's speed have on angle θ?
Justify your answer. 2

(6)

[Turn over

3. A mass of 2·5 kg is attached to a string of negligible mass. The string is wound round a flywheel of radius 0·14 m. A motion sensor, connected to a computer, is placed below the mass as shown in Figure 3A.

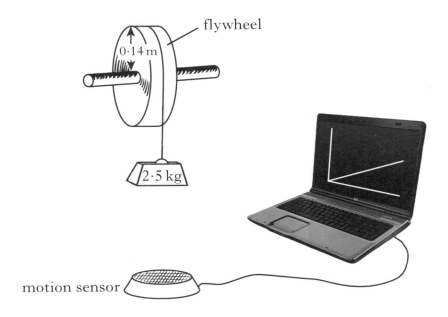

Figure 3A

The mass is released from rest. The computer calculates the linear velocity of the mass as it falls and the angular velocity of the flywheel.

The graph of the angular velocity of the flywheel against time, as displayed on the computer, is shown in Figure 3B.

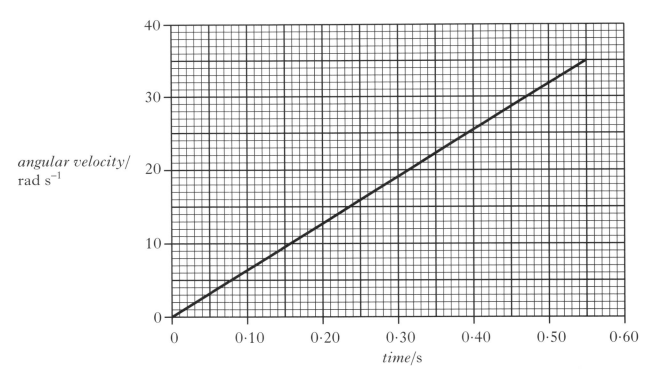

Figure 3B

Marks

3. **(continued)**

(a) Calculate the angular acceleration of the flywheel. **2**

(b) Show that the mass falls a distance of $1 \cdot 3$ m in the first $0 \cdot 55$ seconds. **3**

(c) Use the conservation of energy to calculate the moment of inertia of the flywheel. Assume the frictional force to be negligible. **4**

(9)

[Turn over

Marks

4. (*a*) State the law of conservation of angular momentum.   **1**

(*b*) A student sits on a platform that is free to rotate on a frictionless bearing. The angular velocity of the rotating platform is displayed on a computer.

The student rotates with a hand outstretched at 0·60 m from the axis of rotation as shown in Figure 4A.

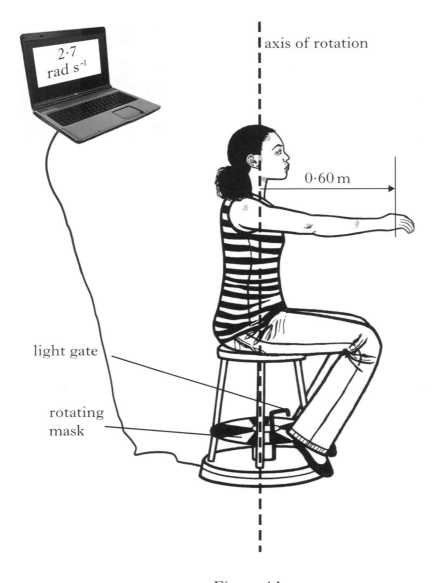

Figure 4A

The moment of inertia of the student and rotating platform is 4·1 kg m². The angular velocity is 2·7 rad s⁻¹.

(i) Calculate the angular momentum of the student and rotating platform.   **2**

4. *(b)* **(continued)**

Marks

(ii) As the student rotates, she grasps a 2·5 kg mass from a stand as shown in Figure 4B.

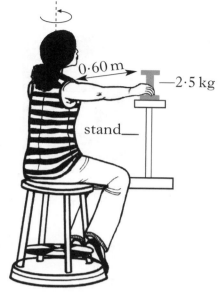

0·60 m

—2·5 kg

stand

Figure 4B

Calculate the angular velocity of the student and platform just after the mass has been grasped.

3

(iii) The student then pulls the mass towards her body.

Explain the effect this has on the angular velocity of the student and the platform.

2

(c) In another investigation the student and platform rotate at 1·5 rad s⁻¹. The student puts one foot on the floor as shown in Figure 4C.

Figure 4C

The frictional force between the student's shoe and the floor brings the student and platform to rest in 0·75 seconds. The new moment of inertia of the student and platform is 4·5 kg m².

Calculate the average frictional torque.

3

(11)

5. A motorised mixer in a DIY store is used to mix different coloured paints.

Paints are placed in a tin and the tin is clamped to the base as shown in Figure 5A.

Figure 5A

The oscillation of the tin in the vertical plane closely approximates to simple harmonic motion.

The amplitude of the oscillation is 0·012 m.

The mass of the tin of paint is 1·4 kg.

Figure 5B shows the graph of the acceleration against displacement for the tin of paint.

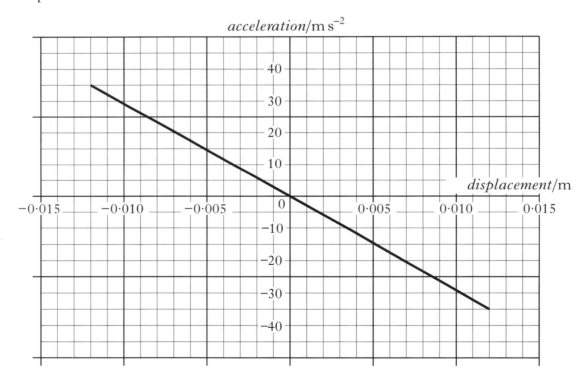

Figure 5B

Marks

5. **(continued)**

   (*a*) Show that the angular frequency ω of the oscillation is $54 \, \text{rad s}^{-1}$.   1

   (*b*) Write an expression for the displacement y of the tin as a function of time. Include appropriate numerical values.   2

   (*c*) Derive an expression for the velocity v of the tin as a function of time. Numerical values should again be included.   2

   (*d*) Calculate the maximum kinetic energy of the tin of paint as it oscillates.   2

                                        (7)

[Turn over

Marks

6. A hollow metal sphere, radius 1·00 mm, carries a charge of −1·92 × 10⁻¹² C.

   (a) Calculate the electric field strength, E, at the surface of the sphere. **2**

   (b) Four students sketch graphs of the variation of electric field strength with distance from the centre of the sphere as shown in Figure 6A.

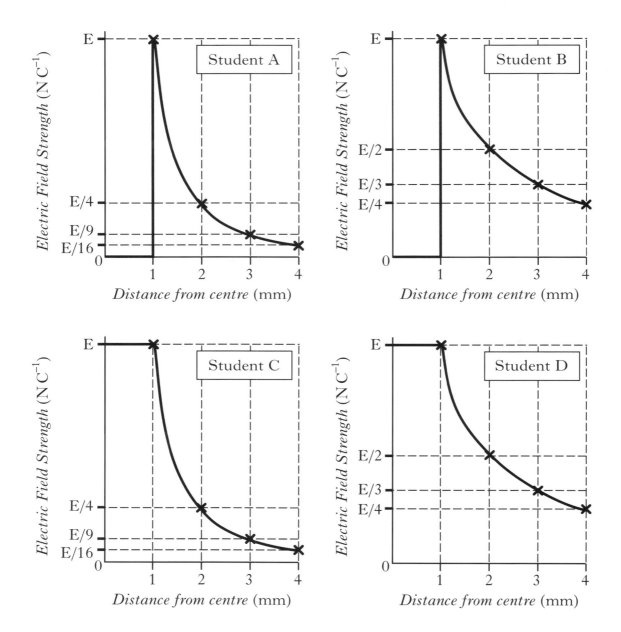

Figure 6A

   (i) Which student has drawn the correct graph? **1**

   (ii) Give **two** reasons to support your choice. **2**

Marks

6. (continued)

(c) Four point charges, Q_1, Q_2, Q_3 and Q_4, each of value $-2\cdot97 \times 10^{-8}\,C$, are held in a square array. The hollow sphere with charge $-1\cdot92 \times 10^{-12}\,C$ is placed $30\cdot0\,mm$ above the centre of the array where it is held stationary by an electrostatic force.

The hollow sphere is $41\cdot2\,mm$ from each of the four charges as shown in Figure 6B.

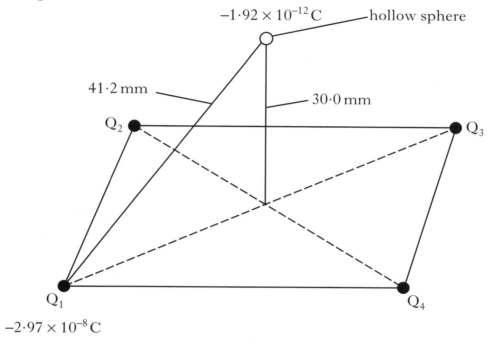

Figure 6B

(i) Calculate the magnitude of the force acting on the sphere due to charge Q_1. 2

(ii) Calculate the vertical component of this force. 2

(iii) Calculate the resultant electrostatic force on the sphere due to the whole array. 1

(iv) Calculate the mass of the sphere. 2

(12)

[Turn over

Marks

7. A beam of protons enters a region of uniform magnetic field, at right angles to the field.

   The protons follow a circular path in the magnetic field until a potential difference is applied across the deflecting plates. The deflected protons hit a copper target. The protons travel through a vacuum. A simplified diagram is shown in Figure 7A.

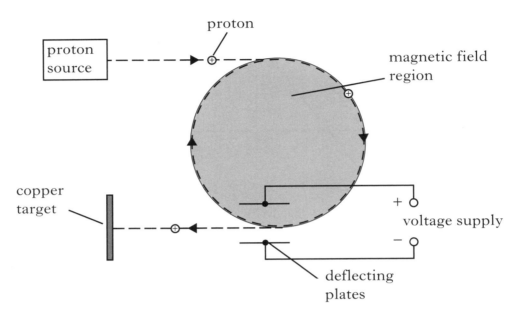

Figure 7A

(a) State the direction of the magnetic field, B. 　　1

(b) The speed of the protons is $6{\cdot}0 \times 10^6\,\mathrm{m\,s^{-1}}$ and the magnetic induction is $0{\cdot}75\,\mathrm{T}$. Calculate the radius of the circular path followed by the protons. 　　3

(c) Calculate the electric field strength required to make the protons move off at a tangent to the circle. 　　2

Marks

7. (continued)

(*d*) A proton of charge q initially travels at speed v directly towards a copper nucleus as shown in Figure 7B. The copper nucleus has charge Q.

copper nucleus

proton

v

Figure 7B

    (i) Show that the distance of closest approach, r, to the copper nucleus is given by

$$\frac{qQ}{2\pi\varepsilon_0 mv^2}.$$

              1

    (ii) Calculate the distance of closest approach for a proton initially travelling at $6\cdot0 \times 10^6\,\mathrm{m\,s^{-1}}$.

              3

    (iii) Name the force that holds the protons together in the copper nucleus.

              1

(*e*) The beam of protons in Figure 7A is replaced by a beam of electrons. The speed of the electrons is the same as the speed of the protons.

State **two** changes that must be made to the magnetic field to allow the electrons to follow the same circular path as the protons.

              2

              (13)

[Turn over

Marks

8. Identification of elements in a semiconductor sample can be carried out using an electron scanner to release atoms from the surface of the sample for analysis. Electrons are accelerated from rest between a cathode and anode by a potential difference of 2·40 kV.

A variable voltage supply connected to the deflection plates enables the beam to scan the sample between points A and B shown in Figure 8A.

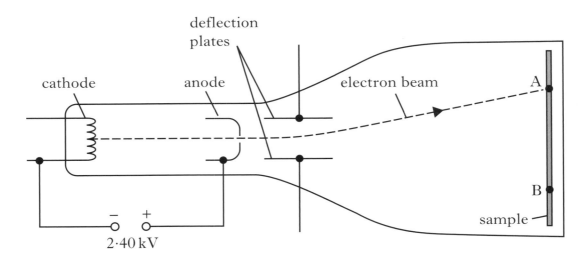

Figure 8A

(a) Calculate the speed of the electrons as they pass through the anode.   2

(b) Explain why the electron beam follows

   (i) a curved path between the plates;   1

   (ii) a straight path beyond the plates.   1

When the potential difference across the deflection plates is 100 V, the electron beam strikes the sample at position A.

(c) The deflection plates are 15·0 mm long and separated by 10·0 mm.

   (i) Show that the vertical acceleration between the plates is $1·76 \times 10^{15}\,\mathrm{m\,s^{-2}}$.   2

   (ii) Calculate the vertical velocity of an electron as it emerges from between the plates.   3

(d) The anode voltage is now increased. State what happens to the length of the sample scanned by the electron beam.

   You must justify your answer.   2

   (11)

Marks

9. A transverse wave travels along a string as shown in Figure 9A.

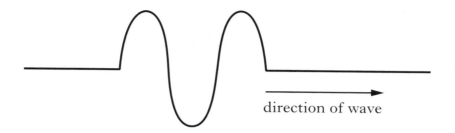

direction of wave

Figure 9A

The equation representing the travelling wave on the string is

$$y = 8{\cdot}6 \times 10^{-2} \sin 2\pi \, (2{\cdot}4t - 2{\cdot}0x)$$

where x and y are in metres and t is in seconds.

(*a*) State the frequency of the wave. **1**

(*b*) Calculate the velocity of the wave. **2**

(*c*) Attached to the end of the string is a very light ring. The ring is free to move up and down a fixed vertical rod.

Figure 9B shows the string after the wave reflects from the vertical rod.

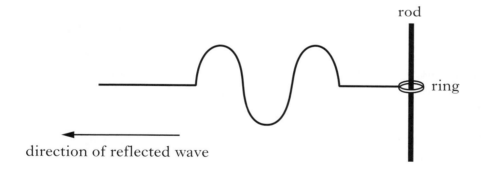

rod

ring

direction of reflected wave

Figure 9B

When the wave reflects, its intensity falls to one quarter of its original value. The frequency and wavelength are constant.

Write the equation that represents this reflected wave. **2**

(5)

[Turn over

Marks

10. (*a*) Explain the formation of coloured fringes when white light illuminates a thin film of oil on a water surface.

2

(*b*) Thin film coatings deposited on glass can be used to make the glass non-reflecting for certain wavelengths of light, as shown in Figure 10A.

The refractive index of the coating is less than glass, but greater than air.

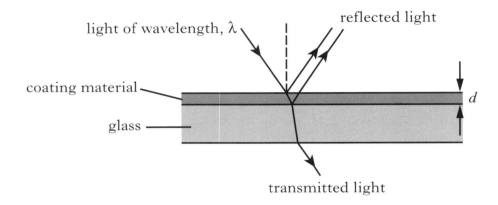

Figure 10A

Show that for near normal incidence

$$d = \frac{\lambda}{4n}$$

where *n* is the refractive index of the coating material and *d* is the thinnest coating that will be non-reflecting for light of wavelength, λ.

2

(*c*) The relationship in (*b*) also applies to radiation of wavelength 780 nm.

A thin film coating has a refractive index of 1·30. For radiation of wavelength 780 nm the minimum thickness for a thin film that is non-reflecting is 0·150 µm. In practice, this thickness is too thin to manufacture.

Calculate the thickness of the next thinnest coating that would be non-reflecting for this wavelength.

2

Marks

10. (continued)

(*d*) Six laser beams provide photons of wavelength 780 nm. These photons collide with rubidium atoms and cause the atoms to come to rest, as shown in Figure 10B.

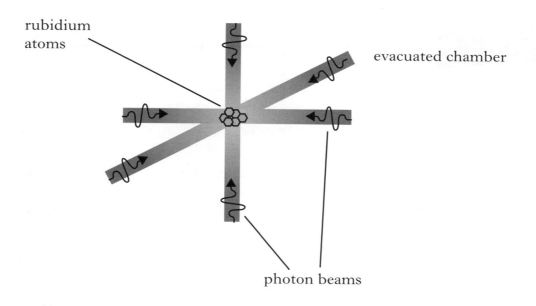

Figure 10B

Each rubidium atom has a mass of $1\cdot43 \times 10^{-25}$ kg and kinetic energy of $4\cdot12 \times 10^{-21}$ J before the lasers are switched on.

(i) Calculate the momentum of a rubidium atom before the lasers are switched on. **3**

(ii) Calculate the momentum of each photon in the laser beams. **2**

(iii) Assuming that all of the momentum of the photons is transferred to a rubidium atom, calculate the number of photons required to bring the atom to rest. **1**

(12)

[Turn over

Marks

11. A light source produces a beam of unpolarised light. The beam of light passes through a polarising filter called a polariser. The transmission axis of the polariser is shown in Figure 11A.

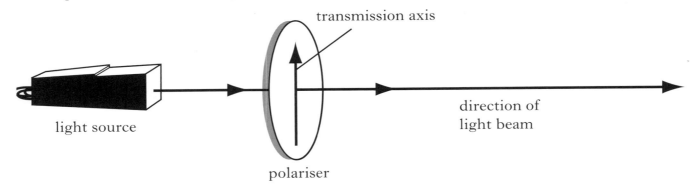

Figure 11A

(a) Explain the difference between the unpolarised light entering the polariser and the plane polarised light leaving the polariser.

1

(b) The plane polarised light passes through a second polarising filter called an analyser.

The irradiance of the light passing through the analyser is measured by a light meter.

The transmission axis of the analyser can be rotated and its angle of rotation measured using a scale as shown in Figure 11B.

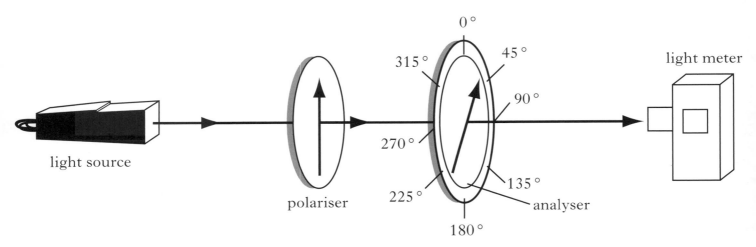

Figure 11B

(i) The analyser is rotated.

State the **two** positions on the analyser scale that will produce a maximum reading of irradiance, I_0, on the light meter.

2

Marks

11. (b) (continued)

(ii) The relationship between the irradiance I detected by the light meter and the angle of rotation θ is given by

$$I = I_0 \cos^2\theta.$$

Explain how the equipment shown in Figure 11B could be used to establish this relationship.

Your answer should include:

- the measurements required;
- a description of how the relationship would be verified.

3

(6)

[END OF QUESTION PAPER]

[BLANK PAGE]

ADVANCED HIGHER

2011

[BLANK PAGE]

X069/701

NATIONAL
QUALIFICATIONS
2011

MONDAY, 23 MAY
1.00 PM – 3.30 PM

PHYSICS
ADVANCED HIGHER

Reference may be made to the Physics Data Booklet.

Answer **all** questions.

Any necessary data may be found in the Data Sheet on page two.

Care should be taken to give an appropriate number of significant figures in the final answers to calculations.

Square-ruled paper (if used) should be placed inside the front cover of the answer book for return to the Scottish Qualifications Authority.

DATA SHEET

COMMON PHYSICAL QUANTITIES

Quantity	Symbol	Value	Quantity	Symbol	Value
Gravitational acceleration on Earth	g	9.8 m s^{-2}	Mass of electron	m_e	$9.11 \times 10^{-31} \text{ kg}$
Radius of Earth	R_E	$6.4 \times 10^6 \text{ m}$	Charge on electron	e	$-1.60 \times 10^{-19} \text{ C}$
Mass of Earth	M_E	$6.0 \times 10^{24} \text{ kg}$	Mass of neutron	m_n	$1.675 \times 10^{-27} \text{ kg}$
Mass of Moon	M_M	$7.3 \times 10^{22} \text{ kg}$	Mass of proton	m_p	$1.673 \times 10^{-27} \text{ kg}$
Radius of Moon	R_M	$1.7 \times 10^6 \text{ m}$	Mass of alpha particle	m_α	$6.645 \times 10^{-27} \text{ kg}$
Mean Radius of Moon Orbit		$3.84 \times 10^8 \text{ m}$	Charge on alpha particle		$3.20 \times 10^{-19} \text{ C}$
Universal constant of gravitation	G	$6.67 \times 10^{-11} \text{ m}^3 \text{ kg}^{-1} \text{ s}^{-2}$	Planck's constant	h	$6.63 \times 10^{-34} \text{ J s}$
Speed of light in vacuum	c	$3.0 \times 10^8 \text{ m s}^{-1}$	Permittivity of free space	ε_0	$8.85 \times 10^{-12} \text{ F m}^{-1}$
Speed of sound in air	v	$3.4 \times 10^2 \text{ m s}^{-1}$	Permeability of free space	μ_0	$4\pi \times 10^{-7} \text{ H m}^{-1}$

REFRACTIVE INDICES

The refractive indices refer to sodium light of wavelength 589 nm and to substances at a temperature of 273 K.

Substance	Refractive index	Substance	Refractive index
Diamond	2·42	Glycerol	1·47
Glass	1·51	Water	1·33
Ice	1·31	Air	1·00
Perspex	1·49	Magnesium Fluoride	1·38

SPECTRAL LINES

Element	Wavelength/nm	Colour	Element	Wavelength/nm	Colour
Hydrogen	656	Red	Cadmium	644	Red
	486	Blue-green		509	Green
	434	Blue-violet		480	Blue
	410	Violet			
	397	Ultraviolet			
	389	Ultraviolet			
Sodium	589	Yellow			

Lasers		
Element	Wavelength/nm	Colour
Carbon dioxide	9550 } 10590 }	Infrared
Helium-neon	633	Red

PROPERTIES OF SELECTED MATERIALS

Substance	Density/ kg m^{-3}	Melting Point/ K	Boiling Point/ K	Specific Heat Capacity/ J kg^{-1} K^{-1}	Specific Latent Heat of Fusion/ J kg^{-1}	Specific Latent Heat of Vaporisation/ J kg^{-1}
Aluminium	2.70×10^3	933	2623	9.02×10^2	3.95×10^5	....
Copper	8.96×10^3	1357	2853	3.86×10^2	2.05×10^5	....
Glass	2.60×10^3	1400	....	6.70×10^2	....	....
Ice	9.20×10^2	273	....	2.10×10^3	3.34×10^5	....
Glycerol	1.26×10^3	291	563	2.43×10^3	1.81×10^5	8.30×10^5
Methanol	7.91×10^2	175	338	2.52×10^3	9.9×10^4	1.12×10^6
Sea Water	1.02×10^3	264	377	3.93×10^3	....	....
Water	1.00×10^3	273	373	4.19×10^3	3.34×10^5	2.26×10^6
Air	1·29	....	....		....	
Hydrogen	9.0×10^{-2}	14	20	1.43×10^4	....	4.50×10^5
Nitrogen	1·25	63	77	1.04×10^3	....	2.00×10^5
Oxygen	1·43	55	90	9.18×10^2	....	2.40×10^5

The gas densities refer to a temperature of 273 K and a pressure of 1.01×10^5 Pa.

Marks

1. In a process called "spallation", protons are accelerated to relativistic speeds and collide with mercury nuclei. Each collision releases neutrons from a mercury nucleus as shown in Figure 1.

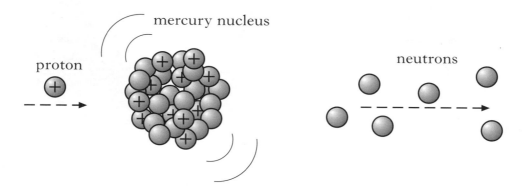

Figure 1

(a) (i) The energy of a proton is $2 \cdot 08 \times 10^{-10}$ J. Calculate the relativistic mass of this proton.   2

   (ii) Calculate the speed of this proton.   2

(b) A neutron produced in the spallation process is slowed to a non-relativistic speed, resulting in a kinetic energy of $3 \cdot 15 \times 10^{-21}$ J.

   (i) Show that the momentum of the neutron is $3 \cdot 25 \times 10^{-24}$ kg m s^{-1}.   2

   (ii) Calculate the de Broglie wavelength of this neutron.   2

(c) In a mercury nucleus, protons experience electrostatic repulsion, yet the nucleus remains stable.

   (i) Name the force responsible for this stability.   1

   (ii) Up to what distance is this force dominant?   1

   (iii) Name the fundamental particles that make up protons and neutrons.   1

   (11)

[Turn over

Marks

2. The front wheel of a racing bike can be considered to consist of 5 spokes and a rim, as shown in Figure 2A.

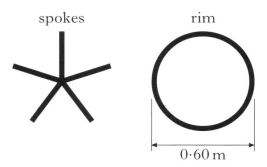

Figure 2A

The mass of each spoke is 0·040 kg and the mass of the rim is 0·24 kg. The wheel has a diameter of 0·60 m.

(a) (i) Each spoke can be considered as a uniform rod. Calculate the moment of inertia of a spoke as the wheel rotates. **2**

(ii) Show that the total moment of inertia of the wheel is $2·8 \times 10^{-2}$ kg m². **2**

(b) The wheel is placed in a test rig and rotated as shown in Figure 2B.

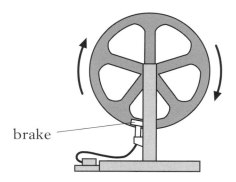

Figure 2B

(i) The tangential velocity of the rim is 19·2 m s⁻¹. Calculate the angular velocity of the wheel. **2**

(ii) The brake is now applied to the rim of the wheel, bringing it uniformly to rest in 6·7 s.

(A) Calculate the angular acceleration of the wheel. **2**

(B) Calculate the torque acting on the wheel. **2**

(10)

Marks

3. An X-ray binary system consists of a star in a **circular** orbit around a black hole as shown in Figure 3A.

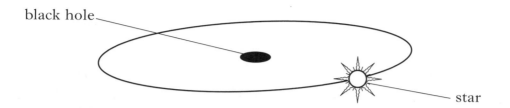

Figure 3A

The star has a mass of $2 \cdot 0 \times 10^{30}$ kg and takes 5·6 days to orbit the black hole. The orbital radius is $3 \cdot 6 \times 10^{10}$ m.

(a) Show that the angular velocity of the star is $1 \cdot 3 \times 10^{-5}$ rad s^{-1}.   **1**

(b) Calculate the mass of the black hole.   **3**

(c) (i) Show that the potential energy of the star in its orbit is $-4 \cdot 4 \times 10^{41}$ J.   **1**

     (ii) Calculate the kinetic energy of the star.   **2**

     (iii) Calculate the total energy of the star due to its motion and position.   **1**

(d) The binary system orbits in the same plane as an earth-based telescope, as shown in Figure 3B.

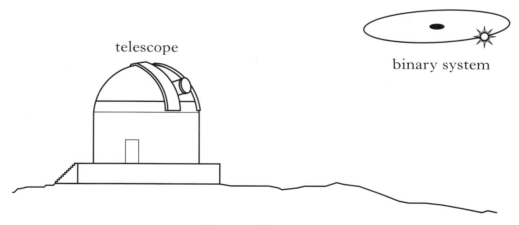

Figure 3B

Light from the star is analysed and found to contain the emission spectrum of hydrogen gas. The frequency of a particular line in the spectrum is monitored and a periodic variation in frequency is recorded.

Explain the periodic variation in the frequency.   **2**

(10)

[Turn over

Marks

4. A design for electrical power generation consists of a large buoy that drives a water column through a turbine as shown in Figure 4.

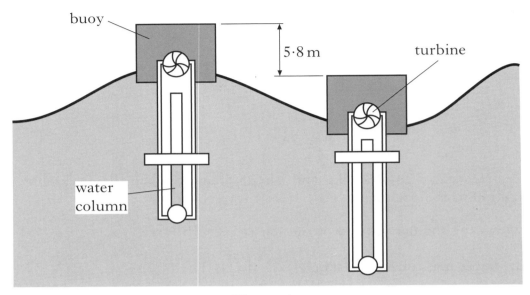

Figure 4

Energy is transferred from the wave motion to the turbines.

The mass of the buoy is 4.0×10^4 kg and its vertical displacement is 5·8 m. The motion of the buoy can be considered to be simple harmonic with a period of oscillation of 5·7 s.

(a) Write an equation that describes the vertical displacement y of the buoy. Numerical values are required.   **2**

(b) Calculate the maximum acceleration of the buoy.   **2**

(c) Where in the motion is the resultant force on the buoy greatest?   **1**

(d) Calculate the maximum kinetic energy of the buoy.   **2**

(e) The water column acts to damp the oscillatory motion of the buoy. How does this affect:

   (i) the period;   **1**

   (ii) the amplitude of the buoy's motion?   **1**

(9)

[Turn over for Question 5 on *Page eight*

Marks

5. A helium-filled metal foil balloon with a radius of 0·35 m is charged by induction. The charge Q on the surface of the balloon is +120 μC. The balloon is considered to be perfectly spherical.

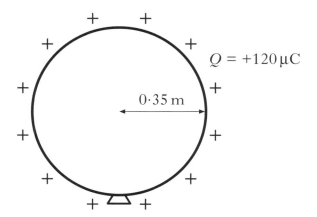

Figure 5A

(a) (i) Using diagrams, or otherwise, describe a procedure to charge the balloon by induction so that an evenly distributed positive charge is left on the balloon.

2

(ii) Calculate the electric field strength at the surface of the balloon.

2

(iii) Sketch a graph of the electric field strength against distance from the centre of the balloon to a point well beyond the balloon's surface. No numerical values are required.

1

(b) Two parallel charged plates are separated by a distance d. The potential difference between the plates is V.

Lines representing the electric field between the plates are shown in Figure 5B.

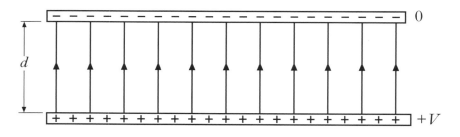

Figure 5B

(i) By considering the work done in moving a point charge q between the plates, derive an expression for the electric field strength E between the plates in terms of V and d.

2

Marks

5. (*b*) (continued)

(ii) The base of a thundercloud is 489 m above an area of open flat ground as shown in Figure 5C.

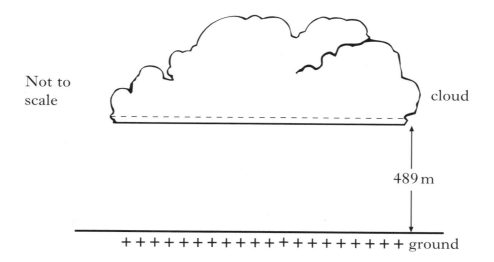

Figure 5C

The uniform electric field strength between the cloud and the ground is $7 \cdot 23 \times 10^4 \, \text{N C}^{-1}$.

Calculate the potential difference between the cloud and the ground.   **1**

(iii) During a lightning strike a charge of $5 \cdot 0 \, \text{C}$ passes between the cloud and the ground in a time of $348 \, \mu\text{s}$. The strike has negligible effect on the potential of the cloud. Calculate the average power of the lightning strike.   **2**

(*c*) An uncharged metal foil balloon is released and floats between the thundercloud and ground, as shown in Figure 5D.

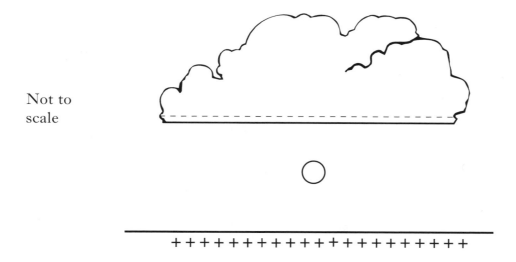

Figure 5D

Draw a diagram showing the charge distribution on the balloon and the resulting electric field around the balloon.   **2**

   (12)

Marks

6. Modern trains have safety systems to ensure that they stop before the end of the line. One system being tested uses a relay operated by a reed switch. The reed switch closes momentarily as it passes over a permanent magnet laid on the track. An inductor in the relay activates the safety system as shown in Figure 6A.

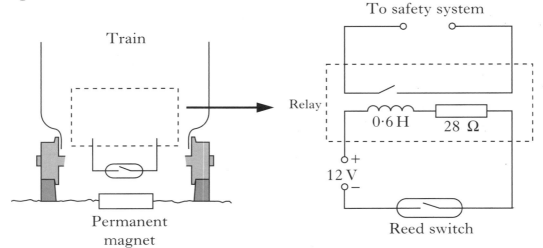

Figure 6A

(a) (i) Explain why there is a short time delay between the reed switch closing and the relay activating.   2

   (ii) The inductor is connected to a 12·0 V d.c. supply. The inductor has an inductance of 0·6 H and the total resistance of the circuit is 28 Ω. Calculate the initial rate of change of current as the reed switch is closed.   2

   (iii) The inductance of the inductor on the train is 0·6 H. Define one henry.   1

   (iv) The reed switch opens as it moves away from the permanent magnet. Explain why a spark occurs across the contacts of the reed switch.   2

   (v) A diode is placed across the inductor to prevent sparks across the reed switch as it opens as shown in Figure 6B. The diode must be chosen to carry the same current as the maximum current which occurs in the circuit when the reed switch closes. Calculate this current.   1

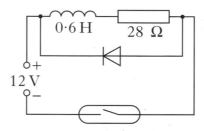

Figure 6B

Marks

6. (continued)

(b) Another safety system prevents trains approaching a stop signal at excessive speed. When a train is travelling too fast the brakes are applied automatically and the train is brought uniformly to rest. An inductor at the front of the train is used to determine the average speed as it travels between the electromagnets 1 and 2 as shown in Figure 6C.

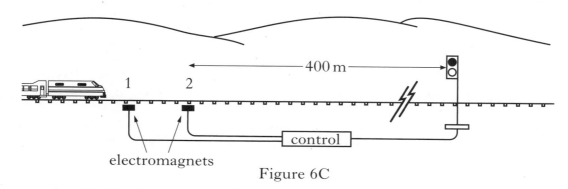

Figure 6C

The train travels between the electromagnets at a constant speed of $99 \, \text{km h}^{-1}$. The brakes are applied automatically as the train passes the second electromagnet. The train is accelerated at $-1 \cdot 0 \, \text{m s}^{-2}$. Show by calculation whether the train stops before the signal.

2

(c) The train is stopped and a passenger hears a siren on another train approaching along a parallel track. The approaching train is travelling at a constant speed of $28 \cdot 0 \, \text{m s}^{-1}$ and the siren produces a sound of frequency $294 \, \text{Hz}$.

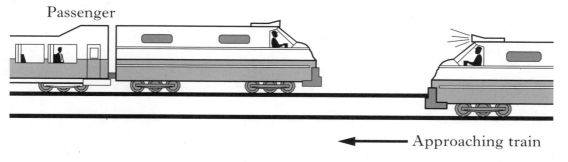

Figure 6D

(i) Show that the frequency f of the sound heard by the passenger is given by

$$f = f_s \left(\frac{v}{v - v_s} \right)$$

where symbols have their usual meaning.

2

(ii) Calculate the frequency of the sound heard by the passenger:

(A) as the train approaches;

1

(B) once the train has passed the passenger.

2

(15)

Marks

7. (*a*) Two very long straight wires X and Y are suspended parallel to each other at a distance *r* apart. The current in X is I_1 and the current in Y is I_2 as shown in Figure 7A.

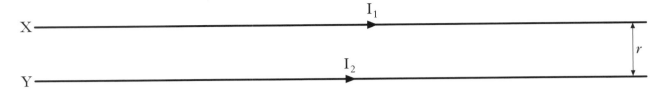

Figure 7A

(i) State the direction of the magnetic force acting on wire X. **Justify your answer**.   2

(ii) The wires are separated by a distance of 360 mm and each wire carries a current of 4·7A. Calculate the force per unit length which acts on each wire.   2

(*b*) A student investigating the force on a current carrying wire placed perpendicular to a uniform magnetic field obtains the following measurements and uncertainties.

Force (N)	0·0058 0·0061 0·0063 0·0057 0·0058 0·0062	
	Scale reading uncertainty	± 1 digit
	Calibration uncertainty	± 0·00005 N
Current (A)	Reading	1·98 A
	Absolute uncertainty	± 0·02 A
Length (m)	Reading	0·054 m
	Absolute uncertainty	± 0·0005 m

(i) From this data, calculate the magnetic induction, B.   3

(ii) Calculate the absolute uncertainty in the value of the force.   3

(iii) Calculate the overall absolute uncertainty in the value of the magnetic induction.   3

(13)

Marks

8. (*a*) Figure 8A shows a current carrying wire of length *l*, perpendicular to a magnetic field *B*. A single charge −*q* moves with constant velocity *v* in the wire. Using the relationship for the force on a current carrying conductor placed in a magnetic field, derive the relationship *F* = *qvB* for the magnitude of the force acting on charge *q*.

2

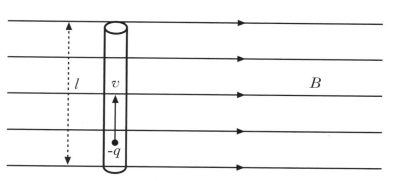

Figure 8A

(*b*) An electron with a speed of $2 \cdot 0 \times 10^6 \, \mathrm{m \, s^{-1}}$ enters a uniform magnetic field at an angle *θ*. The electron follows a helical path as shown in Figure 8B.

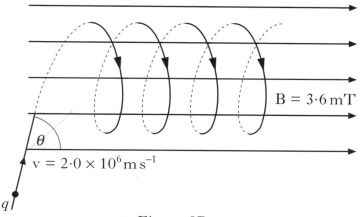

Figure 8B

The uniform magnetic induction is $3 \cdot 6 \, \mathrm{mT}$ and the radius of the helical path is $2 \cdot 8 \, \mathrm{mm}$. Calculate the value of angle *θ*.

3

(*c*) A second electron travelling at the same speed enters the field at a smaller angle *θ*.

Describe how the path of the second electron differs from the first.

2

(7)

[Turn over

Marks

9. A laser-based quality control system to measure thread spacing in fabric samples is being evaluated. The 2-dimensional interference pattern is displayed on a screen shown in Figure 9A.

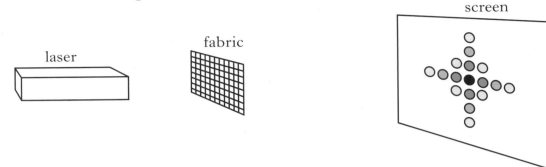

Figure 9A

(a) Explain how this 2-D interference pattern is produced.   2

(b) When a fine beam of laser light of wavelength 488 nm is used, the separation of the maxima in the horizontal direction is 8·00 mm. The distance from the fabric sample to the screen is 3·60 m.

Assume the spaces between the threads act like Young's slits.

Calculate the spacing between the threads in the sample.   2

Marks

9. (continued)

(c) The interference pattern from a standard fabric sample using a 488 nm laser is shown in Figure 9B.

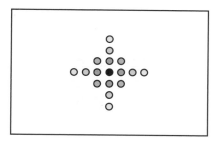

Figure 9B

(i) The 488 nm laser is replaced with a 667 nm laser. Which interference pattern from Figure 9C best represents the new interference pattern? Justify your answer.

2

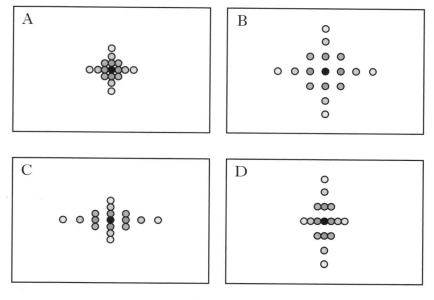

Figure 9C

(ii) The **original** 488 nm laser is restored and the fabric sample is stretched as shown in Figure 9D.

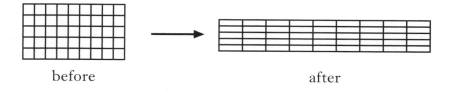

before   after

Figure 9D

Which pattern from Figure 9C best represents the new pattern? Justify your answer.

2

(8)

Marks

10. A stretched wire, supported near its ends, is made to vibrate by touching a tuning fork of unknown frequency to the supporting surface. One of the supports is moved until a stationary wave pattern appears as shown in Figure 10A.

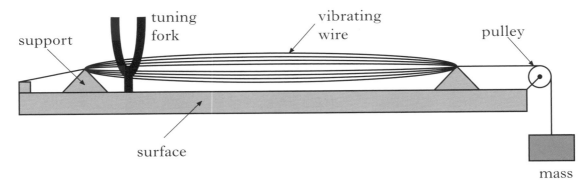

Figure 10A

(a) Explain how waves on this wire produce a stationary wave pattern.   2

(b) The formula for the frequency of the note from a stretched wire is given by:

$$f = \frac{1}{2l}\sqrt{\frac{T}{\mu}}$$

where l is the distance between the supports,
T is the stretching force,
μ is the mass per unit length of the wire.

The results of the experiment are given below:

mass per unit length of wire $= 1.92 \times 10^{-4} \, \text{kg m}^{-1}$
distance between the supports $= 0.780 \, \text{m}$
mass of load on wire $= 4.02 \, \text{kg}$

(i) The table below gives information about the note produced by tuning forks of different frequency. Identify the note most likely to correspond to the tuning fork used in the experiment.   2

Note	A	B	C	D	E	F	G
Frequency (Hz)	220	245	262	294	330	349	392

Marks

10. (b) (continued)

(ii) A second tuning fork produces the pattern shown in Figure 10B. Suggest a frequency for this tuning fork.

1

(5)

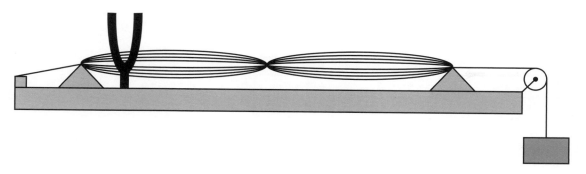

Figure 10B

[*END OF QUESTION PAPER*]

[BLANK PAGE]

ADVANCED HIGHER

2012

[BLANK PAGE]

X069/13/01

NATIONAL
QUALIFICATIONS
2012

MONDAY, 28 MAY
1.00 PM – 3.30 PM

PHYSICS
ADVANCED HIGHER

Reference may be made to the Physics Data Booklet.

Answer **all** questions.

Any necessary data may be found in the Data Sheet on *Page two*.

Care should be taken to give an appropriate number of significant figures in the final answers to calculations.

Square-ruled paper (if used) should be placed inside the front cover of the answer book for return to the Scottish Qualifications Authority.

DATA SHEET
COMMON PHYSICAL QUANTITIES

Quantity	Symbol	Value	Quantity	Symbol	Value
Gravitational acceleration on Earth	g	$9.8\,\text{m s}^{-2}$	Mass of electron	m_p	$9.11 \times 10^{-31}\,\text{kg}$
Radius of Earth	R_E	$6.4 \times 10^6\,\text{m}$	Charge on electron	e	$-1.60 \times 10^{-19}\,\text{C}$
Mass of Earth	M_E	$6.0 \times 10^{24}\,\text{kg}$	Mass of neutron	m_n	$1.675 \times 10^{-27}\,\text{kg}$
Mass of Moon	M_M	$7.3 \times 10^6\,\text{kg}$	Mass of proton	m_p	$1.673 \times 10^{-27}\,\text{kg}$
Radius of Moon	R_M	$1.7 \times 10^6\,\text{m}$	Mass of alpha particle	m_α	$6.645 \times 10^{-27}\,\text{kg}$
Mean Radius of Moon Orbit		$3.84 \times 10^8\,\text{m}$	Charge on alpha particle		$3.20 \times 10^{-19}\,\text{C}$
Universal constant of gravitation	G	$6.67 \times 10^{-11}\,\text{m}^3\,\text{kg}^{-1}\,\text{s}^{-2}$	Planck's constant	h	$6.63 \times 10^{-34}\,\text{J s}$
Speed of light in vacuum	c	$3.0 \times 10^8\,\text{m s}^{-1}$	Permittivity of free space	ε_0	$8.85 \times 10^{-12}\,\text{F m}^{-1}$
Speed of sound in air	v	$3.4 \times 10^2\,\text{m s}^{-1}$	Permeability of free space	μ_0	$4\pi \times 10^{-7}\,\text{H m}^{-1}$

REFRACTIVE INDICES
The refractive indices refer to sodium light of wavelength 589 nm and to substances at a temperature of 273 K.

Substance	Refractive index	Substance	Refractive index
Diamond	2·42	Glycerol	1·47
Glass	1·51	Water	1·33
Ice	1·31	Air	1·00
Perspex	1·49	Magnesium Fluoride	1·38

SPECTRAL LINES

Element	Wavelength/nm	Colour	Element	Wavelength/nm	Colour
Hydrogen	656	Red	Cadmium	644	Red
	486	Blue-green		509	Green
	434	Blue-violet		480	Blue
	410	Violet	Lasers		
	397	Ultraviolet	Element	Wavelength/nm	Colour
	389	Ultraviolet	Carbon dioxide	9550 } 10590 }	Infrared
Sodium	589	Yellow	Helium-neon	633	Red

PROPERTIES OF SELECTED MATERIALS

Substance	Density/ kg m^{-3}	Melting Point/ K	Boiling Point/K	Specific Heat Capacity/ J kg^{-1} K^{-1}	Specific Latent Heat of Fusion/ J kg^{-1}	Specific Latent Heat of Vaporisation/ J kg^{-1}
Aluminium	2.70×10^3	933	2623	9.02×10^2	3.95×10^5	. . . .
Copper	8.96×10^3	1357	2853	3.86×10^2	2.05×10^5	. . . .
Glass	2.60×10^3	1400	. . . .	6.70×10^2	. . . .	. . . .
Ice	9.20×10^2	273	. . . .	2.10×10^3	3.34×10^5	. . . .
Glycerol	1.26×10^3	291	563	2.43×10^3	1.81×10^5	8.30×10^5
Methanol	7.91×10^2	175	338	2.52×10^3	9.9×10^4	1.12×10^6
Sea Water	1.02×10^3	264	377	3.93×10^3	. . . .	. . . .
Water	1.00×10^3	273	373	4.19×10^3	3.34×10^5	2.26×10^6
Air	1·29	. . . .	. . . .	. . . .	. . . .	. . . .
Hydrogen	9.0×10^{-2}	14	20	1.43×10^4	. . . .	4.50×10^5
Nitrogen	1·25	63	77	1.04×10^3	. . . .	2.00×10^5
Oxygen	1·43	55	90	9.18×10^2	. . . .	2.40×10^4

The gas densities refer to a temperature of 273 K and a pressure of 1.01×10^5 Pa.

Marks

1. (*a*) A beta particle travelling at high speed has a relativistic mass 1·8 times its rest mass.

    (i) Calculate the speed of the beta particle.   **2**

    (ii) Calculate the relativistic energy of the beta particle at this speed.   **2**

    (iii) Name the force associated with beta decay.   **1**

(*b*) Electrons exhibit both wave-like and particle-like behaviour.

    (i) Give **one** example of experimental evidence which suggests an electron exhibits wave-like behaviour.   **1**

    (ii) Give **one** example of experimental evidence which suggests an electron exhibits particle-like behaviour.   **1**

         (7)

[Turn over

Marks

2. (a) The acceleration of a particle moving in a straight line is given by

$$a = \frac{dv}{dt}$$

where the symbols have their usual meaning.

   (i) Show, by integration, that when a is constant

$$v = u + at.$$

2

   (ii) Show that when a is constant

$$v^2 = u^2 + 2as.$$

1

(b) The path taken by a short track speed skater is shown in Figure 2A. The path consists of two straights each of length 29·8 m and two semicircles each of radius 8·20 m.

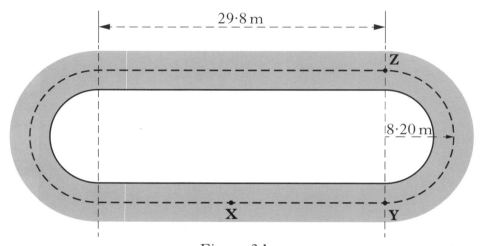

Figure 2A

Starting at point **X**, half way along the straight, the skater accelerates uniformly from rest. She reaches a speed of 9·64 m s⁻¹ at point **Y**, the end of the straight.

   (i) Calculate the acceleration of the skater.

2

   (ii) The skater exits the curve at point **Z** with a speed of 10·9 m s⁻¹.

      Calculate the average **angular** acceleration of the skater between **Y** and **Z**.

3

Marks

2. (continued)

(*c*) When this speedskater reaches a curve she leans inwards and digs the blade of the skate into the ice as shown in Figure 2B. Force F indicates the reaction of the ice on the skater.

Figure 2B

    (i) Explain how force F allows her to maintain a curved path.

2

    (ii) The skater approaches the next curve at a greater speed and slides off the track. Explain, **in terms of forces**, why this happens.

1

(11)

[Turn over

Marks

3. To test a springboard a diver takes up a position at the end of the board and sets up an oscillation as shown in Figure 3A. The oscillation approximates to simple harmonic motion. The board oscillates with a frequency of 0·76 Hz. The end of the board moves through a vertical distance of 0·36 m.

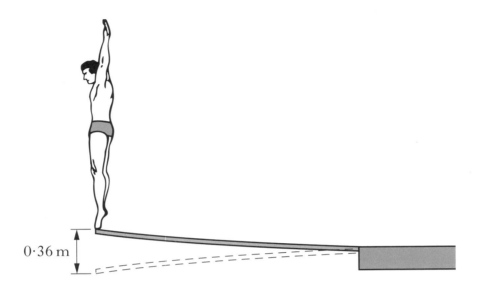

0·36 m

Figure 3A

(a) (i) Write an expression for the vertical displacement y of the end of the board as a function of time t. Include appropriate numerical values.   **2**

(ii) The diver increases the amplitude of the oscillation. The frequency remains constant. Show that the amplitude when the diver just loses contact with the board is 0·43 m.   **1**

3. (continued)

Marks

(b) A sport scientist analyses a dive. At one point during the dive, shown in Figure 3B, he approximates the diver's body to be two rods of equal mass rotating about point G. One rod has a length of 0·94 m the other of 0·90 m. The diver has a mass of 66·0 kg

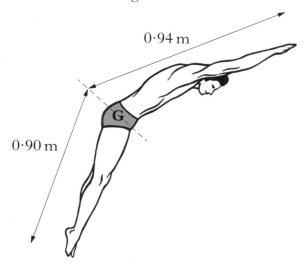

Figure 3B

(i) Calculate the approximate moment of inertia of the diver. **3**

(ii) The diver's true moment of inertia about point G is found to be 10·25 kg m². Account for any difference between the value calculated in part (i) and the true value. **1**

(iii) In the position shown in Figure 3B, the diver has an initial angular velocity of 0·55 rad s⁻¹. He changes his position to that shown in Figure 3C. The diver now has a moment of inertia of 7·65 kg m². Calculate the angular velocity of the diver in this new position. **2**

Figure 3C

(c) (i) Calculate the change in rotational kinetic energy between these two positions. **2**

(ii) Account for the difference in rotational kinetic energy. **1**

(12)

Marks

4. (a) Show that the gravitational field strength at the surface of Pluto, mass M_p, is given by

$$g = \frac{GM_p}{r^2}$$

where the symbols have their usual meanings.

1

(b) Figure 4A shows how the gravitational potential varies with distance from the centre of Pluto.

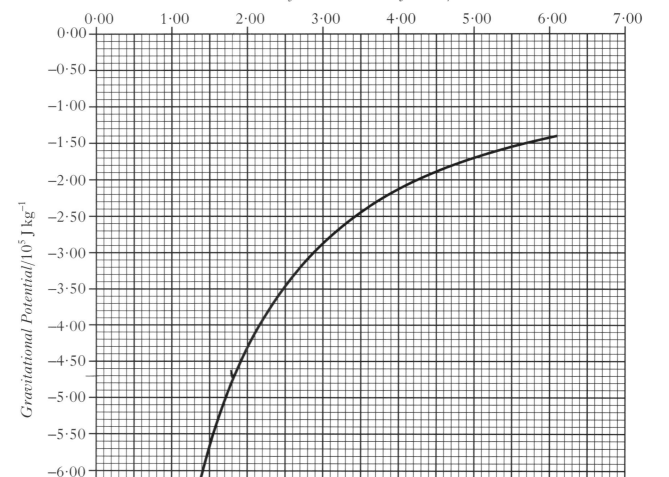

Figure 4A

(i) The mass of Pluto is $1 \cdot 27 \times 10^{22}$ kg. Calculate the gravitational field strength at the surface of Pluto.

2

(ii) A meteorite hits the surface of Pluto and ejects a lump of ice of mass 112 kg. The ice is captured in an orbit $1 \cdot 80 \times 10^6$ m from the centre of Pluto. Calculate the gravitational potential energy of the ice at this height.

2

4. (continued)

Marks

(c) In 2015 the New Horizons space probe is due to arrive at Pluto. The space probe will move between Pluto and its moon, Charon, as shown in Figure 4B. Pluto has a mass seven times that of Charon and their average separation is $1 \cdot 96 \times 10^7$ m.

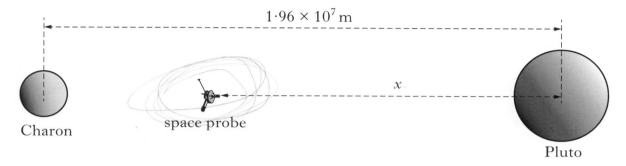

Not to scale

Figure 4B

Calculate the distance x from the centre of Pluto where the resultant gravitational force acting on the probe is zero. Ignore any orbital motion of the two objects.

3

(8)

[Turn over

Marks

5. (a) An uncharged conducting sphere is suspended from a fixed point X by an insulating thread of negligible mass as shown in Figure 5A. A charged plate is then placed close to the sphere as shown in Figure 5B.

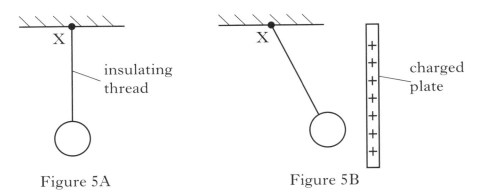

Figure 5A   Figure 5B

Explain why the uncharged sphere is attracted to the charged plate. You may use a diagram to help explain your answer.   **1**

(b) The sphere is now given a negative charge of 140 nC and placed between a pair of parallel plates with a separation of 42 mm as shown in Figure 5C.

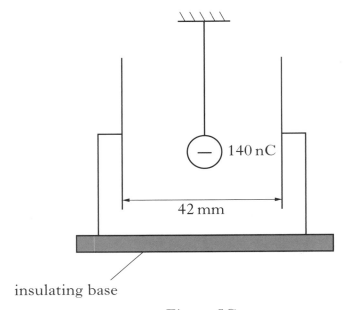

Figure 5C

5. (*b*) (continued)   *Marks*

When a potential difference is applied to the plates the sphere is deflected through an angle θ as shown in Figure 5D.

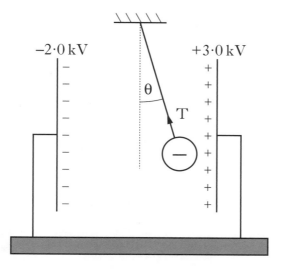

Figure 5D

   (i)   Calculate the electric field strength between the plates.   **2**

   (ii) Calculate the electrostatic force acting on the sphere due to the electric field.   **2**

  (iii) The mass of the sphere is $4 \cdot 0 \times 10^{-3}$ kg.

        Calculate the magnitude and direction of the tension T in the supporting thread.   **3**

(*c*) The plates are now moved a short distance to the right without touching the sphere. The distance between the plates is unchanged.

     Does the angle θ increase, decrease or stay the same? Justify your answer.   **2**

                                             (10)

[Turn over

Marks

6. A positively charged particle travelling at $2 \cdot 29 \times 10^6 \, \mathrm{m\,s^{-1}}$ enters a magnetic field of uniform magnetic induction $2 \cdot 50 \, \mathrm{T}$ as shown in Figure 6A.

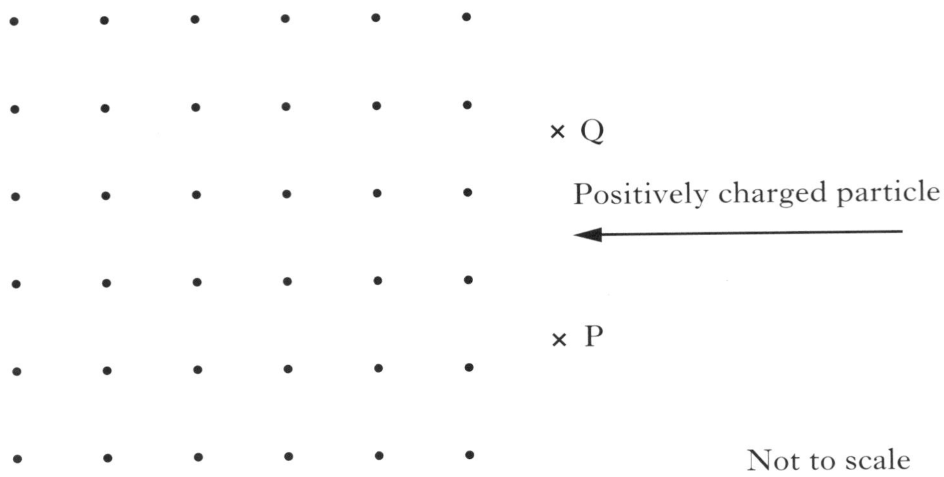

Figure 6A

The direction of the magnetic field is out of the page. The particle follows a semicircular path before exiting the field.

(a) (i)   State whether the particle will exit the field at point P or point Q.   1

    (ii) Show that the charge to mass ratio of the particle is given by

$$\frac{q}{m} = \frac{v}{rB}$$

        where the symbols have their usual meaning.   1

   (iii) The radius of the path taken by the particle is $19 \cdot 0 \, \mathrm{mm}$.

        Use information from the data sheet to identify the charged particle. You must justify your answer by calculation.   3

   (iv) Calculate the time between the particle entering and leaving the magnetic field.   2

   (v)   An identical particle travelling at twice the speed of the original particle enters the field at the same point.

        How does the time spent in the magnetic field by this particle compare with the original? Justify your answer.   2

Marks

6. (continued)

(*b*) An unknown particle also travelling at $2\cdot29 \times 10^6\,\text{m s}^{-1}$ enters the field as shown in Figure 6B. The path taken by this particle is shown.

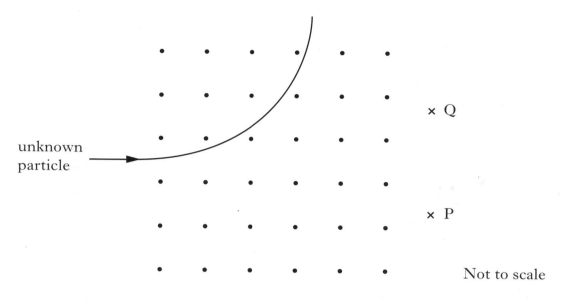

Figure 6B

What can you conclude about:

(i) the charge of the unknown particle;   **1**

(ii) the charge to mass ratio of the unknown particle?   **1**

(11)

[Turn over

Marks

7. Precision inductors can be produced using laser technology.

    A thin film of copper is deposited on a ceramic core. A carbon dioxide laser is then used to cut the copper to form a coil as shown in Figure 7A.

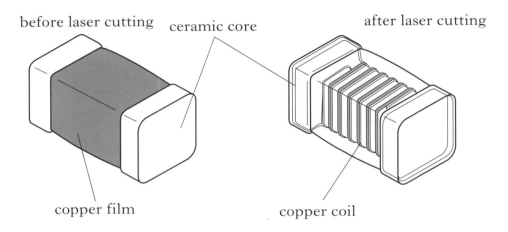

before laser cutting   ceramic core   after laser cutting

copper film   copper coil

Figure 7A

(a) Each photon from the laser has a momentum of $6{\cdot}26 \times 10^{-29}\,\text{kg m s}^{-1}$.

    (i)   Calculate the wavelength of each photon.   **2**

    (ii) The inductor has an inductance of $0{\cdot}1\,\text{H}$.

          Explain what is meant by an inductance of $0{\cdot}1\,\text{H}$.   **1**

(b) The rate of change of current for a **different inductor** is investigated using a datalogger as shown in Figure 7B. This inductor has inductance L and a resistance of $2\,\Omega$.

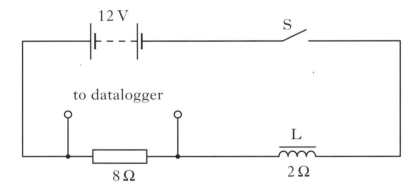

12 V

S

to datalogger

L

$8\,\Omega$ $2\,\Omega$

Figure 7B

The graph shown in Figure 7C shows how the rate of change of current dI/dt in the circuit varies with time from the instant switch S is closed.

7. (b) (continued)

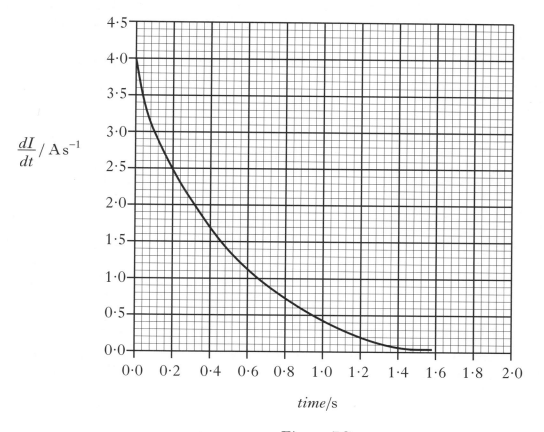

Figure 7C

(i) Describe what happens to the magnetic field strength associated with the inductor between 0 and 1·6 s.

1

(ii) Use information from the graph to determine the inductance L.

2

(iii) Sketch a graph to show how the voltage across the 8 Ω resistor varies during this time. Numerical values are required on both axes.

2

(iv) Calculate the maximum energy stored by the inductor in this circuit.

2

[Turn over

7. (continued)   *Marks*

(c) A student sets up the circuit shown in Figure 7D to investigate the relationship between current and frequency for a capacitor and for an inductor.

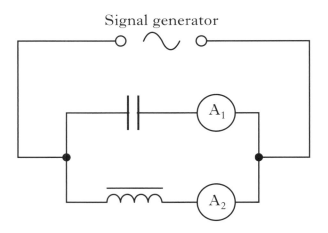

Figure 7D

At frequency of 75 Hz the readings on A_1 and A_2 are the same.

Explain what happens to the readings on each ammeter as the frequency is increased from 75 Hz to 150 Hz.

Assume that the supply voltage remains constant.   **2**

(12)

Marks

8. A long thin horizontal conductor AB carrying a current of 25 A is supported by two fine threads of negligible mass. The tension in each supporting thread is T as shown in Figure 8A.

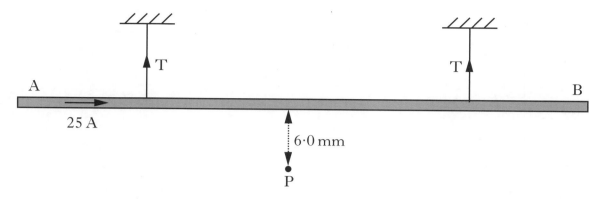

Figure 8A

(a) Calculate the magnetic induction at a point P, 6·0 mm directly below conductor AB.

2

(b) A second conductor CD carrying current I is now fixed in a position a distance r directly below AB as shown in Figure 8B. CD is unable to move.

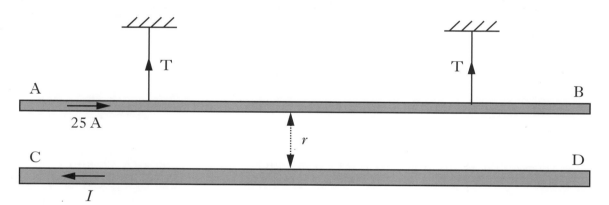

Figure 8B

(i) Explain why there is a force of repulsion between conductors AB and CD.

2

(ii) Show that the force per unit length acting on each conductor can be written as

$$\frac{F}{l} = \frac{5 \cdot 0 \times 10^{-6} I}{r}.$$

1

(iii) The mass per unit length of the conductor AB is $5 \cdot 70 \times 10^{-3} \, \text{kg m}^{-1}$. When the conductors are separated by 6·0 mm, the current I in conductor CD is gradually increased. Calculate the value of I which reduces the tension in the supporting threads to zero.

3

(8)

Marks

9. A travelling wave moves from left to right at a speed of $1 \cdot 25 \, \text{m s}^{-1}$.

Figure 9A represents this wave at a time *t*. P and Q are particles on the wave.

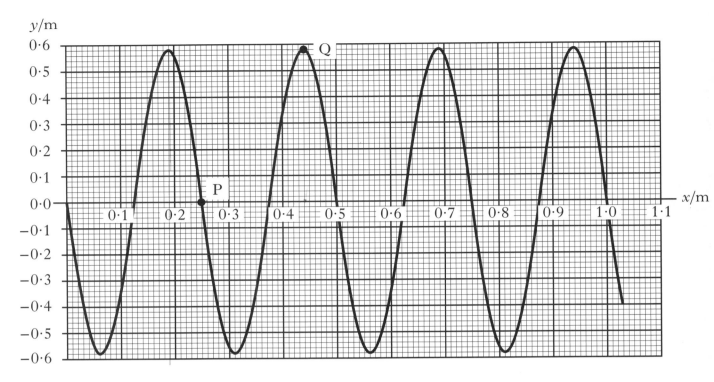

Figure 9A

(a) (i)   Determine the wavelength of the wave.   1

     (ii) State the amplitude of the wave.   1

     (iii) Calculate the frequency of the wave.   1

     (iv) What is the phase difference, in radians, between particles P and Q?   2

(b) Write an equation for this travelling wave in terms of *y*, *x* and *t*.

     Numerical values are required.   2

(c) State the equation for a wave of half the amplitude travelling in the opposite
     direction.   1

                                                                                                 (8)

Marks

10. Figure 10 shows a model helicopter flying in a straight horizontal path from student A to student B.

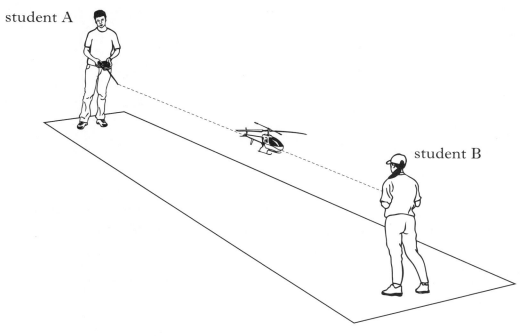

student A

student B

Figure 10

The helicopter has a siren that emits sound of frequency 595 Hz.

(a) For the first two seconds the displacement of the helicopter relative to student A is described by the equation

$$s = 4 \cdot 1\, t^2.$$

   (i) Calculate the velocity of the helicopter when $t = 2 \cdot 0$ s.　　　2

   (ii) Suggest what happens to the frequency of the sound heard by student B as the helicopter accelerates towards her.

      Justify your answer.　　　2

(b) After $2 \cdot 0$ s the helicopter continues towards student B with a constant velocity. Calculate the frequency of the sound heard by student B.　　　2

(6)

[Turn over

11. A student uses laser light of wavelength of 529 nm to determine the separation of the slits in a Young's double slit arrangement as shown in Figure 11A.

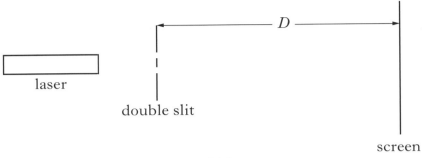

Figure 11A

A pattern of bright green dots is observed on the screen. The distance between the central maximum and the next bright dot is Δx as shown in Figure 11B.

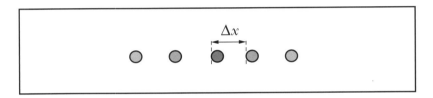

Figure 11B

The distance Δx is measured with a metre stick. The distance D is measured with a tape measure.

The screen is moved and Δx and D are measured for six positions of the screen. Each pair of measurements is repeated five times.

The student uses the results to plot the graph shown in Figure 11C.

Marks

11. (continued)

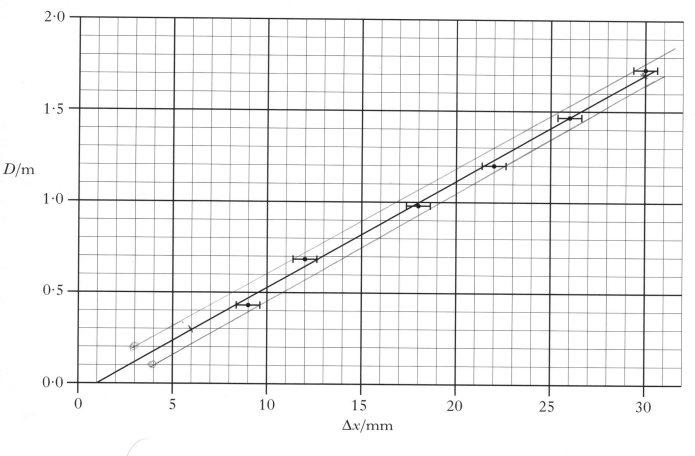

Figure 11C

(a) (i) Using the gradient of the graph, calculate the separation of the double slits.   2

(ii) Suggest a reason why no error bars are shown for the slit to screen distance, D.   1

(iii) Other than repeating the measurements, suggest **two** improvements to the student's experimental technique.   2

(b) The experiment is repeated using a very bright LED in place of the laser. The LED emits light in the wavelength range 535 to 555 nm.

Other than a slight colour change, state **two** differences in the pattern observed on the screen compared to the pattern shown in Figure 11B.   2

(7)

[*END OF QUESTION PAPER*]

[BLANK PAGE]

[BLANK PAGE]

X069/13/01

NATIONAL
QUALIFICATIONS
2013

MONDAY, 27 MAY
1.00 PM – 3.30 PM

PHYSICS
ADVANCED HIGHER

Reference may be made to the Physics Data Booklet.

Answer **all** questions.

Any necessary data may be found in the Data Sheet on *Page two*.

Care should be taken to give an appropriate number of significant figures in the final answers to calculations.

Square-ruled paper (if used) should be placed inside the front cover of the answer book for return to the Scottish Qualifications Authority.

DATA SHEET
COMMON PHYSICAL QUANTITIES

Quantity	Symbol	Value	Quantity	Symbol	Value
Gravitational acceleration on Earth	g	$9.8\ \mathrm{m\,s^{-2}}$	Mass of electron	m_e	$9.11 \times 10^{-31}\ \mathrm{kg}$
Radius of Earth	R_E	$6.4 \times 10^6\ \mathrm{m}$	Charge on electron	e	$-1.60 \times 10^{-19}\ \mathrm{C}$
Mass of Earth	M_E	$6.0 \times 10^{24}\ \mathrm{kg}$	Mass of neutron	m_n	$1.675 \times 10^{-27}\ \mathrm{kg}$
Mass of Moon	M_M	$7.3 \times 10^{22}\ \mathrm{kg}$	Mass of proton	m_p	$1.673 \times 10^{-27}\ \mathrm{kg}$
Radius of Moon	R_M	$1.7 \times 10^6\ \mathrm{m}$	Mass of alpha particle	m_α	$6.645 \times 10^{-27}\ \mathrm{kg}$
Mean Radius of Moon Orbit		$3.84 \times 10^8\ \mathrm{m}$	Charge on alpha particle		$3.20 \times 10^{-19}\ \mathrm{C}$
Universal constant of gravitation	G	$6.67 \times 10^{-11}\ \mathrm{m^3\,kg^{-1}\,s^{-2}}$	Planck's constant	h	$6.63 \times 10^{-34}\ \mathrm{J\,s}$
			Permittivity of free space	ε_0	$8.85 \times 10^{-12}\ \mathrm{F\,m^{-1}}$
Speed of light in vacuum	c	$3.0 \times 10^8\ \mathrm{m\,s^{-1}}$	Permeability of free space	μ_0	$4\pi \times 10^{-7}\ \mathrm{H\,m^{-1}}$
Speed of sound in air	v	$3.4 \times 10^2\ \mathrm{m\,s^{-1}}$			

REFRACTIVE INDICES

The refractive indices refer to sodium light of wavelength 589 nm and to substances at a temperature of 273 K.

Substance	Refractive index	Substance	Refractive index
Diamond	2·42	Glycerol	1·47
Glass	1·51	Water	1·33
Ice	1·31	Air	1·00
Perspex	1·49	Magnesium Fluoride	1·38

SPECTRAL LINES

Element	Wavelength/nm	Colour	Element	Wavelength/nm	Colour
Hydrogen	656	Red	Cadmium	644	Red
	486	Blue-green		509	Green
	434	Blue-violet		480	Blue
	410	Violet			
	397	Ultraviolet		*Lasers*	
	389	Ultraviolet	Element	Wavelength/nm	Colour
			Carbon dioxide	9550 ⎱ 10590 ⎰	Infrared
Sodium	589	Yellow	Helium-neon	633	Red

PROPERTIES OF SELECTED MATERIALS

Substance	Density/ kg m^{-3}	Melting Point/ K	Boiling Point/K	Specific Heat Capacity/ J kg^{-1} K^{-1}	Specific Latent Heat of Fusion/ J kg^{-1}	Specific Latent Heat of Vaporisation/ J kg^{-1}
Aluminium	2.70×10^3	933	2623	9.02×10^2	3.95×10^5	. . . .
Copper	8.96×10^3	1357	2853	3.86×10^2	2.05×10^5	. . . .
Glass	2.60×10^3	1400	. . . .	6.70×10^2	. . . .	. . . .
Ice	9.20×10^2	273	. . . .	2.10×10^3	3.34×10^5	. . . .
Glycerol	1.26×10^3	291	563	2.43×10^3	1.81×10^5	8.30×10^5
Methanol	7.91×10^2	175	338	2.52×10^3	9.9×10^4	1.12×10^6
Sea Water	1.02×10^3	264	377	3.93×10^3	. . . .	. . . .
Water	1.00×10^3	273	373	4.19×10^3	3.34×10^5	2.26×10^6
Air	1·29	. . . .	. . . .	. . . .	. . . .	. . . .
Hydrogen	9.0×10^{-2}	14	20	1.43×10^4	. . . .	4.50×10^5
Nitrogen	1·25	63	77	1.04×10^3	. . . .	2.00×10^5
Oxygen	1·43	55	90	9.18×10^2	. . . .	2.40×10^4

The gas densities refer to a temperature of 273 K and a pressure of 1.01×10^5 Pa.

Marks

1. A stunt driver is attempting to "loop the loop" in a car as shown in Figure 1. Before entering the loop the car accelerates along a horizontal track.

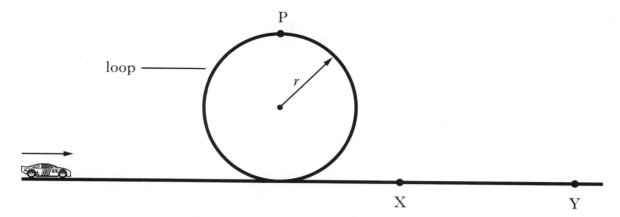

Figure 1

The radius *r* of the circular loop is 6·2 m.
The total mass of the car and driver is 870 kg.

(a) Show that the car must have a minimum speed of 7·8 m s^{-1} at point P to avoid losing contact with the track.

2

(b) During one attempt the car is moving at a speed of 9·0 m s^{-1} at point P.

(i) Draw a labelled diagram showing the vertical forces acting on the car at point P.

1

(ii) Calculate the size of each force.

3

(c) When the car exits the loop the driver starts braking at point X. For one particular run the displacement of the car from point X **until the car comes to rest** at point Y is given by the equation

$$s = 9·1t - 3·2\,t^2$$

Sketch a graph to show how the displacement of the car varies with time between points X and Y.

Numerical values are required on both axes.

3

(9)

[Turn over

Marks

2. The entrance to a building is through a revolving system consisting of 4 doors that rotate around a central axis as shown in Figure 2A.

Figure 2A

The moment of inertia of the system about the axis of rotation is $54\,kg\,m^2$. When it rotates, a constant frictional torque of $25\,N\,m$ acts on the system.

(a) The system is initially stationary. On entering the building a person exerts a constant force F perpendicular to a door at a distance of $1.2\,m$ from the axis of rotation as shown in Figure 2B.

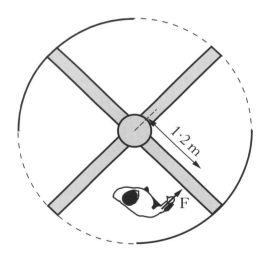

Figure 2B

The angular acceleration of the system is $2.4\,rad\,s^{-2}$.

   (i) Calculate the magnitude of the applied force F.   3

   (ii) The applied force is removed and the system comes to rest in $3.6\,s$. Calculate the angular displacement of the door during this time.   3

Marks

2. (continued)

(*b*) On exiting the building the person exerts the same magnitude of force F on a door at the same distance from the axis of rotation.

The force is now applied as shown in Figure 2C.

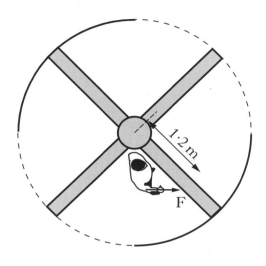

Figure 2C

How does the angular acceleration of the door system compare to that given in part (*a*)?

Justify your answer.

2

(8)

[Turn over

Marks

3. Planets outside our solar system are called exoplanets.

   One exoplanet moves in a circular orbit around a star as shown in Figure 3.

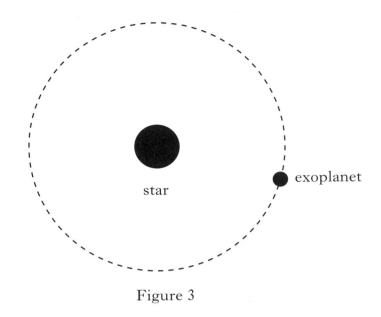

Figure 3

The period of orbit is 14 days. The mass M_s of the star is $1 \cdot 7 \times 10^{30}$ kg.

(a) (i) Show that the radius of the orbit can be given by the relationship

$$r^3 = GM_s \frac{T^2}{4\pi^2}$$

where the symbols have their usual meaning.   2

   (ii) Calculate the radius of this orbit.   1

(b) The radius of the exoplanet is $1 \cdot 2 \times 10^8$ m and its mass is $5 \cdot 4 \times 10^{26}$ kg. Calculate the value of the gravitational field strength g on the surface of the exoplanet.   2

(c) Astrophysicists have identified many black holes in the universe.

   (i) State what is meant by the term *black hole*.   1

   (ii) A newly discovered object has a mass of $4 \cdot 2 \times 10^{30}$ kg and a radius of $2 \cdot 6 \times 10^4$ m.

   Show by calculation whether or not this object is a black hole.   2

(8)

Marks

4. A "saucer" swing consists of a bowl shaped seat of mass 1·2 kg suspended by four ropes of negligible mass as shown in Figure 4A.

Figure 4A

When the empty seat is pulled back slightly from its rest position and released, its motion approximates to simple harmonic motion.

(a) Define the term *simple harmonic motion*. 1

(b) The acceleration-time graph for the seat with no energy loss is shown in Figure 4B.

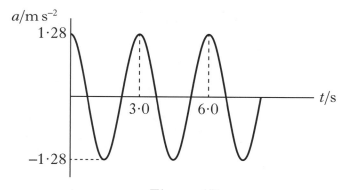

Figure 4B

   (i) Show that the amplitude of the motion is 0·29 m. 3

   (ii) Calculate the velocity of the seat when its displacement is 0·10 m. 2

(c) Calculate the displacement of the seat when the kinetic energy and potential energy are equal. 3

(9)

[Turn over

Marks

5.   The Bohr model of the atom suggests that the angular momentum of an electron orbiting a nucleus is quantised.

A hydrogen atom consists of a single electron orbiting a single proton. Figure 5 shows some of the possible orbits for the electron in a hydrogen atom.

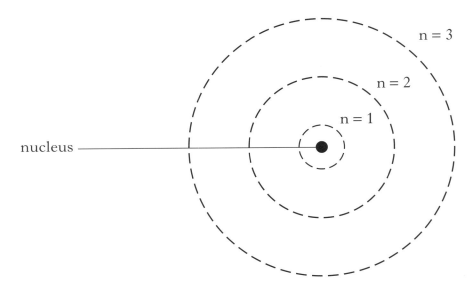

Figure 5

The table shows the values of the radii for the first three orbits.

Orbit number, n	Orbital radius/10^{-10} m
1	0·53
2	2·1
3	4·8

(a)   (i)   Calculate the speed of the electron in orbit number 3.   **2**

(ii)   Calculate the de Broglie wavelength associated with this electron.   **2**

(iii)   What is the name given to the branch of physics that treats electrons as waves and predicts their position in terms of probability?   **1**

(b)   Compare the magnitudes of the electrostatic and gravitational forces between an electron in orbit number 1 and the proton in the nucleus.

Justify your answer by calculation.   **3**

(8)

[Turn over for Question 6 on *Page ten*

Marks

6. A research physicist is investigating collisions between protons and the nuclei of metallic elements. Protons are accelerated **from rest** across a potential difference of 4·0 MV. The protons move through a vacuum and collide with a metal target as shown in Figure 6A.

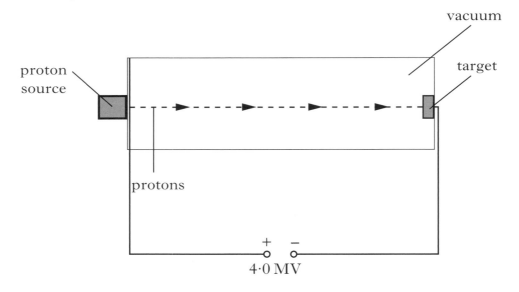

Figure 6A

(a) (i) Calculate the maximum speed of the protons as they hit the target.   **2**

    (ii) In one test the researcher uses zirconium as the target. A proton of charge q and velocity v travels directly towards a zirconium nucleus as shown in Figure 6B. The zirconium nucleus has charge Q.

proton   zirconium nucleus

Figure 6B

Show that the distance of closest approach r to the metal target is given by

$$r = \frac{qQ}{2\pi\varepsilon_{o}mv^2}$$

where the symbols have their usual meaning.   **1**

    (iii) Calculate the distance of closest approach for a proton travelling towards a zirconium nucleus in the target.   **3**

Marks

6. **(continued)**

(b) At CERN protons are accelerated to speeds approaching the speed of light. Calculate the relativistic energy of a proton moving at $0.8c$.   **4**

(c) A student visiting CERN asks why the protons in the nucleus of an atom do not just fly apart. Explain **fully** why protons in a nucleus do not behave in this way.   **3**

(13)

[Turn over

Marks

7. In a nuclear power station liquid sodium is used to cool parts of the reactor. An electromagnetic pump keeps the coolant circulating. The sodium enters a perpendicular magnetic field and an electric current, I, passes through it. A force is experienced by the sodium causing it to flow in the direction shown in Figure 7A.

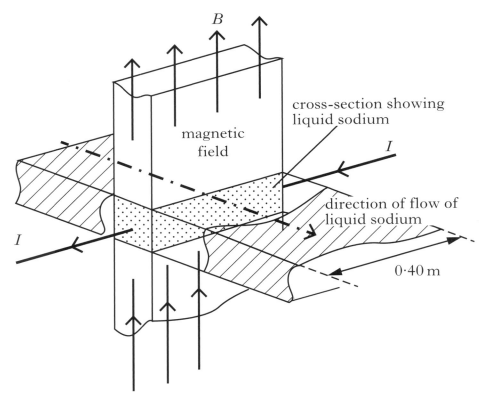

Figure 7A

The magnetic induction B is 0.20 T. The current I in the sodium is 2.5 A and is perpendicular to the magnetic field.

1

(a) Define one tesla.

(b) Calculate the force acting on the 0.40 m length of sodium within the magnetic field.

2

(c) The pump is moved during maintenance and as a result the direction of the magnetic field is changed so that it is no longer perpendicular to the current. What effect does this have on the rate of flow of sodium passing through the pump?

You **must** justify your answer.

2

Marks

7. (continued)

(*d*) An engineer must install a long, straight, current carrying wire AB close to the pump and is concerned that the magnetic induction produced may interfere with the safe working of the pump.

The wire is 750 mm from the pump and carries a current of 0·60 A.

Show by calculation that the magnetic induction at this distance is negligible.

2

(*e*) A second long straight wire CD is installed parallel to the first wire AB as shown in Figure 7B.

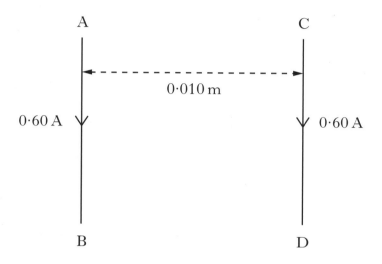

Figure 7B

(i) It also carries a current of 0·60 A in the same direction as in the first wire AB. Calculate the size of the force per unit length exerted on wire CD by wire AB.

2

(ii) State the direction of the force on the wire CD.

Justify your answer.

2

(11)

[Turn over

Marks

8. In 1909 Robert Millikan devised an experiment to investigate the charge on a small oil drop. Using a variable power supply he adjusted the potential difference between two horizontal parallel metal plates until an oil drop was held stationary between them as shown in Figure 8.

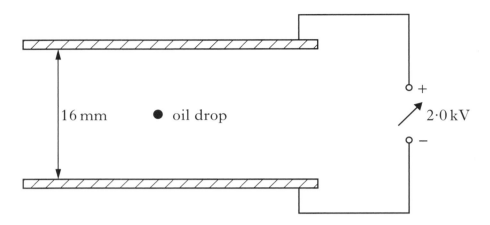

Figure 8

(a) What was Millikan's main conclusion from this experiment?   1

(b) Draw a labelled diagram showing the forces acting on the stationary oil drop.   1

(c) The parallel plates are fixed 16 mm apart. In one experiment the charge on the oil drop was found to be $2 \cdot 4 \times 10^{-18}$ C.

Calculate the mass of the oil drop.   3

                                                                                        (5)

9. The charge Q on a hollow metal sphere is $(-15 \cdot 0 \pm 0 \cdot 4)\,\mu C$. The sphere has a radius r of $(0 \cdot 65 \pm 0 \cdot 02)\,m$.

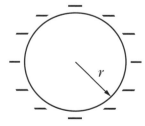

Figure 9

(a) Calculate the electrostatic potential at the surface of the metal sphere.   2

(b) Calculate the absolute uncertainty in the electrostatic potential.   2

(c) State the electrostatic potential at the centre of the sphere.   1

                                                                                        (5)

Marks

10. A 0·40 H inductor of negligible resistance is connected in a circuit as shown in Figure 10. Switch S is initially open.

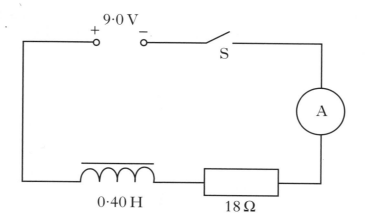

Figure 10

  (a) (i) The switch S is closed. Sketch a graph of current against time giving numerical values on the current axis.   2

       (ii) Explain fully the shape of the graph.   2

  (b) Calculate the initial rate of change of current when switch S is closed.   2

                                                     (6)

[Turn over

Marks

11. High quality *optical flats* made from glass are often used to test components of optical instruments. A high quality optical flat has a very smooth and flat surface.

(a) During the manufacture of an optical flat, the quality of the surface is tested by placing it on top of a high quality flat. This results in a thin air wedge between the flats as shown in Figure 11A.

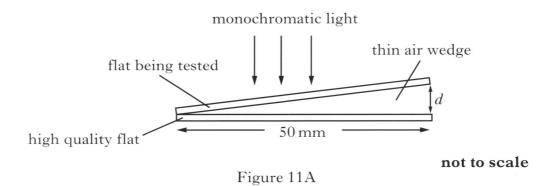

Figure 11A

The thickness d of the air wedge is $6 \cdot 2 \times 10^{-5}$ m.

Monochromatic light is used to illuminate the flats from above. When viewed from above using a travelling microscope, a series of interference fringes is observed as shown in Figure 11B.

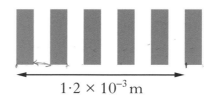

Figure 11B

Calculate the wavelength of the monochromatic light.   3

(b) A second flat is tested using the same method as in part (a). This flat is slightly curved as shown in Figure 11C.

Figure 11C

Draw the fringe pattern observed.   1

Marks

11. (continued)

(*c*) Good quality optical flats often have a non-reflecting coating of magnesium fluoride applied to the surface as shown in Figure 11D.

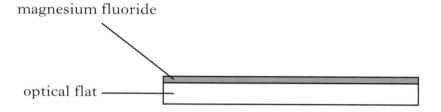

magnesium fluoride

optical flat

Figure 11D

(i) With the aid of a diagram explain fully how the coating reduces reflections from the flat for monochromatic light.   **2**

(ii) Calculate the minimum thickness of magnesium fluoride required to make the flat non-reflecting for yellow light from a sodium lamp.   **2**

(8)

[Turn over

Marks

12. A water wave of frequency 2·5 Hz travels from left to right.

Figure 12 represents the displacement *y* of the water at one instant in time.

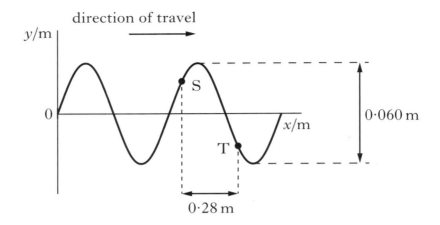

Figure 12

Points S and T are separated by a horizontal distance of 0·28 m.

The phase difference between these two points is 3·5 radians.

(*a*) Calculate the wavelength of this wave.　　　　　　　　　　　　　　　　2

(*b*) A second wave with double the frequency travels in the same direction through the water. This wave has five times the intensity of the wave in part (*a*).

Calculate:

   (i) the speed of this wave;　　　　　　　　　　　　　　　　　　　　1

   (ii) the amplitude of this wave.　　　　　　　　　　　　　　　　　　2

(5)

Marks

13. A student is investigating polarisation of waves.

    (a) State what is meant by *plane polarised light*.   1

    (b) The student wishes to investigate polarisation of sound waves and asks a teacher for suitable apparatus. The teacher says that sound waves cannot be polarised.

    Why can sound waves not be polarised?   1

    (c) (i) While doing some background reading the student discovers that the Brewster angle i_p for the liquid solvent triethylamine is given as 54·5°. Explain using a diagram what is meant by the Brewster angle.   2

        (ii) Calculate the refractive index of triethylamine.   1

                                                                                         (5)

[END OF QUESTION PAPER]

[BLANK PAGE]

[BLANK PAGE]

X069/13/01

NATIONAL QUALIFICATIONS 2014	THURSDAY, 22 MAY 1.00 PM – 3.30 PM	PHYSICS ADVANCED HIGHER

Reference may be made to the Physics Data Booklet.

Answer **all** questions.

Any necessary data may be found in the Data Sheet on *Page two*.

Care should be taken to give an appropriate number of significant figures in the final answers to calculations.

Square-ruled paper (if used) should be placed inside the front cover of the answer book for return to the Scottish Qualifications Authority.

DATA SHEET

COMMON PHYSICAL QUANTITIES (APPROXIMATE)

Quantity	Symbol	Value	Quantity	Symbol	Value
Gravitational acceleration on Earth	g	9.8 m s^{-2}	Mass of electron	m_e	$9.11 \times 10^{-31} \text{ kg}$
Radius of Earth	R_E	$6.4 \times 10^6 \text{ m}$	Charge on electron	e	$-1.60 \times 10^{-19} \text{ C}$
Mass of Earth	M_E	$6.0 \times 10^{24} \text{ kg}$	Mass of neutron	m_n	$1.675 \times 10^{-27} \text{ kg}$
Mass of Moon	M_M	$7.3 \times 10^{22} \text{ kg}$	Mass of proton	m_p	$1.673 \times 10^{-27} \text{ kg}$
Radius of Moon	R_M	$1.7 \times 10^6 \text{ m}$	Mass of alpha particle	m_α	$6.645 \times 10^{-27} \text{ kg}$
Mean Radius of Moon Orbit		$3.84 \times 10^8 \text{ m}$	Charge on alpha particle		$3.20 \times 10^{-19} \text{ C}$
Universal constant of gravitation	G	$6.67 \times 10^{-11} \text{ m}^3 \text{ kg}^{-1} \text{ s}^{-2}$	Planck's constant	h	$6.63 \times 10^{-34} \text{ J s}$
Speed of light in vacuum $\frac{1}{\sqrt{\mu_0 \varepsilon_0}}$	c	$3.0 \times 10^8 \text{ m s}^{-1}$ $2.997956377 \times 10^8 \text{ m s}^{-1}$	Permittivity of free space	ε_0	$8.854 \times 10^{-12} \text{ F m}^{-1}$
Speed of sound in air	v	$3.4 \times 10^2 \text{ m s}^{-1}$	Permeability of free space	μ_0	$4\pi \times 10^{-7} \text{ H m}^{-1}$

REFRACTIVE INDICES

The refractive indices refer to sodium light of wavelength 589 nm and to substances at a temperature of 273 K.

Substance	Refractive index	Substance	Refractive index
Diamond	2·42	Glycerol	1·47
Glass	1·51	Water	1·33
Ice	1·31	Air	1·00
Perspex	1·49	Magnesium Fluoride	1·38

SPECTRAL LINES

Element	Wavelength/nm	Colour	Element	Wavelength/nm	Colour
Hydrogen	656	Red	Cadmium	644	Red
	486	Blue-green		509	Green
	434	Blue-violet		480	Blue
	410	Violet		Lasers	
	397	Ultraviolet	Element	Wavelength/nm	Colour
	389	Ultraviolet	Carbon dioxide	9550 } 10590 }	Infrared
Sodium	589	Yellow	Helium-neon	633	Red

PROPERTIES OF SELECTED MATERIALS

Substance	Density/ kg m^{-3}	Melting Point/ K	Boiling Point/K	Specific Heat Capacity/ J kg^{-1} K^{-1}	Specific Latent Heat of Fusion/ J kg^{-1}	Specific Latent Heat of Vaporisation/ J kg^{-1}
Aluminium	2.70×10^3	933	2623	9.02×10^2	3.95×10^5	. . . .
Copper	8.96×10^3	1357	2853	3.86×10^2	2.05×10^5	. . . .
Glass	2.60×10^3	1400	. . . .	6.70×10^2	. . . .	. . . .
Ice	9.20×10^2	273	. . . .	2.10×10^3	3.34×10^5	. . . .
Glycerol	1.26×10^3	291	563	2.43×10^3	1.81×10^5	8.30×10^5
Methanol	7.91×10^2	175	338	2.52×10^3	9.9×10^4	1.12×10^6
Sea Water	1.02×10^3	264	377	3.93×10^3	. . . .	. . . .
Water	1.00×10^3	273	373	4.19×10^3	3.34×10^5	2.26×10^6
Air	1·29	. . . .	. . . .	. . . .	. . . .	. . . .
Hydrogen	9.0×10^{-2}	14	20	1.43×10^4	. . . .	4.50×10^5
Nitrogen	1·25	63	77	1.04×10^3	. . . .	2.00×10^5
Oxygen	1·43	55	90	9.18×10^2	. . . .	2.40×10^4

The gas densities refer to a temperature of 273 K and a pressure of 1.01×10^5 Pa.

Marks

1. An HS1000 tidal powered electricity generator is installed on the sea bed close to Orkney. The generator is shown in Figure 1.

Figure 1

The blades of the turbine are 7·8 m long. The moving tidal water spins the twin turbines to generate electricity. The maximum power output of the HS1000 is 1·0 MW.

(a) (i) When operating, the tangential velocity of the tip of the blades is 8·2 m s^{-1}.

   Calculate the angular velocity of the blades.   **2**

   (ii) When the blades are rotating at this angular velocity, the rotational kinetic energy is 100 M J.

   Calculate the total moment of inertia of the rotating parts of the HS1000.   **2**

   (iii) State **one** assumption made in the calculation(s).   **1**

(b) Starting from rest it takes 42 seconds for the turbine to reach the angular velocity calculated in (a)(i).

   Calculate the torque acting on the turbines.   **3**

   (8)

[Turn over

Marks

2. A student uses two methods to calculate the moment of inertia of a solid cylinder about its central axis.

   (a) In the first method the student measures the mass of the cylinder to be 0·115 kg and the diameter to be 0·030 m.

   Calculate the moment of inertia of the cylinder.   **2**

   (b) In a second method the student allows the cylinder to roll down a slope and measures the final speed at the bottom of the slope to be 1·60 m s⁻¹. The cylinder has a diameter of 0·030 m and the slope has a height of 0·25 m, as shown in Figure 2.

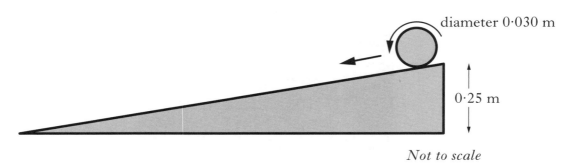

diameter 0·030 m

0·25 m

Not to scale

Figure 2

   Using the conservation of energy, calculate the moment of inertia.   **4**

   (c) Explain why the moment of inertia found in part (b) is greater than in part (a).   **1**

   (7)

Marks

3. A team of astrophysicists from a Scottish University has discovered, orbiting a nearby star, an exoplanet with the same mass as Earth.

By considering the escape velocity of the exoplanet, the composition of its atmosphere can be predicted.

(a) (i) Explain the term *escape velocity*.   1

   (ii) Derive the expression for escape velocity in terms of the exoplanet's mass and radius.   2

  (iii) The radius of this exoplanet is 1·7 times that of the Earth.

Calculate the escape velocity of the exoplanet.   3

(b) Astrophysicists consider that a gas will be lost from the atmosphere of a planet if the typical molecular velocity (v_{rms}) is $\frac{1}{6}$ or more of the escape velocity for that planet.

The table below gives v_{rms} for selected gases at 273 K.

Gas	$v_{rms}(\text{m s}^{-1})$
Hydrogen	1838
Helium	1845
Nitrogen	493
Oxygen	461
Methane	644
Carbon dioxide	393

The atmospheric temperature of this exoplanet is 273 K.

Predict which of these gases could be found in its atmosphere.   2

         (8)

[Turn over

Marks

4. Car engines use the ignition of fuel to release energy which moves the pistons up and down, causing the crankshaft to rotate.

The vertical motion of the piston approximates to simple harmonic motion.

Figure 4 shows different positions of a piston in a car engine.

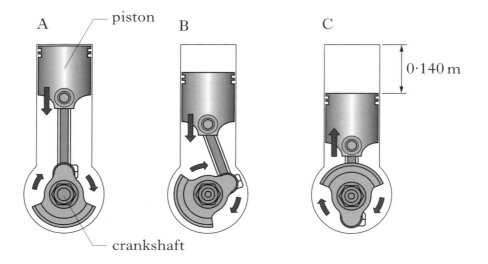

Figure 4

(a) Define *simple harmonic motion*.   1

(b) Determine the amplitude of the motion.   1

(c) In this engine the crankshaft rotates at 1500 revolutions per minute and the piston has a total mass of 1·40 kg.

    (i)   Calculate the maximum acceleration of the piston.   3

    (ii) Calculate the maximum kinetic energy of the piston.   2

    (7)

Marks

5. In 1928, Davisson and Germer fired a beam of electrons through a very thin layer of nickel in a vacuum, which resulted in the production of a diffraction pattern.

(a) (i)   What did they conclude from the results of their experiment?   1

(ii) Give **one** example of experimental evidence that photons of light exhibit particle properties.   1

(b) Calculate the de Broglie wavelength of an electron travelling at $4 \cdot 4 \times 10^6 \, \text{m s}^{-1}$.   2

(c) A 20 g bullet travelling at $300 \, \text{m s}^{-1}$ passes through a 500 mm gap in a target.

Using the data given, explain why no diffraction pattern is observed.   2

(d) (i)   Describe the Bohr model of the hydrogen atom.   2

(ii) Calculate the angular momentum of an electron in the third stable orbit of a hydrogen atom.   2

(10)

[Turn over

Marks

6.   Four point charges P, Q, R and S are fixed in a rectangular array. Point charge A is placed at the centre of the rectangle, 2·5 mm from each of the fixed charges, as shown in Figure 6A.

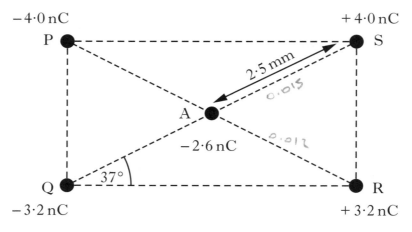

Figure 6A

(a)   (i)   Calculate the magnitude of the force exerted on charge A due to charge P.

2

(ii) The magnitude of the force exerted on charge A due to charge Q is 0·012 N.

Calculate the **resultant force** exerted on charge A due to all the fixed charges.

3

(b)   The four fixed charges are removed and charge A is now fixed. Position B is at a distance *r* from charge A as shown in Figure 6B.

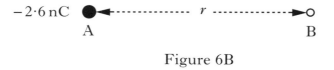

Figure 6B

(i)   The electrostatic potential at B is −37 V.

Calculate the distance *r*.

2

Marks

6. (b) (continued)

(ii) A charge may be moved from B to position C along two possible paths, 1 and 2, shown in Figure 6C.

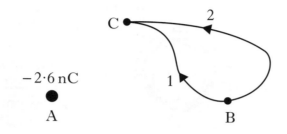

Figure 6C

Compare the work done in moving the charge by the two different routes.

1

(8)

[Turn over

Marks

7.   A student investigates the behaviour of electrons in electric fields. In one experiment the student uses the equipment shown in Figure 7A to investigate the charge to mass ratio for an electron.

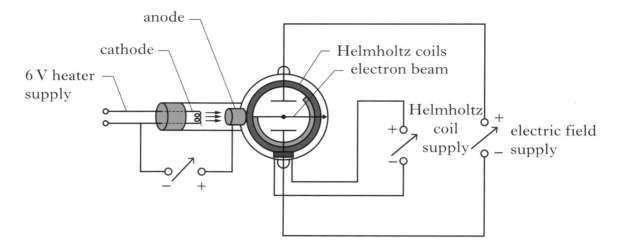

Figure 7A

(a) An electron of mass m and charge q is accelerated from rest between a cathode and anode by a potential difference V. Electron velocity v is measured using the interaction between electric and magnetic fields.

   (i) Show that the charge to mass ratio q/m is given by

$$\frac{q}{m} = \frac{v^2}{2V}.$$ 1

   (ii) The student obtained the following data.

Electron velocity ($\times 10^7 \, \mathrm{m\,s^{-1}}$)	Potential difference (kV)
2·92	2·48
2·73	2·15
2·61	2·00
2·47	1·75
2·26	1·56

   (A) The student decides to calculate q/m individually for each pair of results. Calculate the values obtained.   2

   (B) The student then suggests that the mean of these values is the best estimate of q/m. Calculate this mean value.   1

   (C) Another student correctly explains that this is an inappropriate method for calculating q/m and that a more appropriate method is to draw a graph.

   Explain in detail how a graphical approach could be used to determine a value for q/m.   2

Marks

7. (continued)

(b) In a second experiment the potential difference between cathode and anode is set at 2·08 kV as shown in Figure 7B.

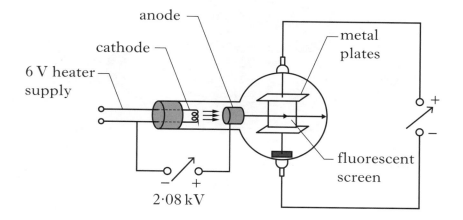

Figure 7B

(i) Show that the speed of an electron on reaching the anode is $2\cdot70 \times 10^7\,\text{m s}^{-1}$.

1

(ii) The electron then passes through the uniform electric field produced between parallel plates. The path of this electron between the parallel plates is shown in Figure 7C.

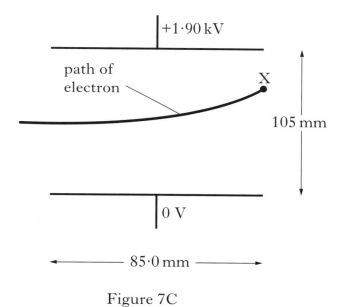

Figure 7C

Calculate the vertical displacement of the electron at point X.

3

(iii) The potential difference between the cathode and the anode is now increased to 2·40 kV.

Will electrons exit the field at the same position, higher or lower than point X as shown in Figure 7C. Justify your answer.

2

(12)

Mark.

8. Research is currently being carried out into nuclear fusion as a future source of energy. One approach uses a magnetic field to contain ionised gas, known as plasma, in a hollow doughnut-shaped ring. A simplified design is shown in Figure 8A.

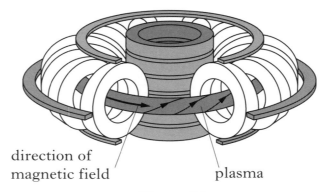

direction of
magnetic field   plasma

Figure 8A

The motion of a charged gas particle is determined by the angle θ between its velocity v and the magnetic induction B as shown in Figure 8B.

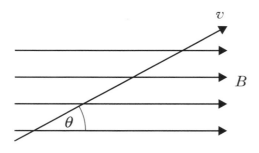

Figure 8B

(a) In the case where $\theta = 90°$ the particles undergo circular motion, perpendicular to the magnetic field.

   (i) Show that for a charged particle of mass m, charge q and velocity v in a field of magnetic induction B the radius of rotation is given by

$$r = \frac{mv}{qB}$$

1

Marks

8. (*a*) (continued)

   (ii) A deuterium ion is moving with a velocity of $2 \cdot 4 \times 10^7 \, \text{m s}^{-1}$ perpendicular to the magnetic field. The maximum diameter of the circular motion, permitted by the design, is $0 \cdot 50 \, \text{m}$.

   Properties of ions present in the plasma are given in the table below.

Ion	Symbol	Mass ($\times 10^{-27}$ kg)	Charge ($\times 10^{-19}$ C)
Hydrogen	H^+	$1 \cdot 686$	$1 \cdot 60$
Deuterium	D^+	$3 \cdot 343$	$1 \cdot 60$
Tritium	T^+	$5 \cdot 046$	$1 \cdot 60$

   Calculate the magnetic induction B required to constrain the deuterium ion within the maximum permitted diameter.   2

   (iii) Calculate the maximum period of rotation for this deuterium ion.   2

(*b*) Another deuterium ion is travelling at $2 \cdot 4 \times 10^7 \, \text{m s}^{-1}$ at an angle of $40°$ to the direction of magnetic induction. This results in the ion undergoing helical motion as shown in Figure 8C.

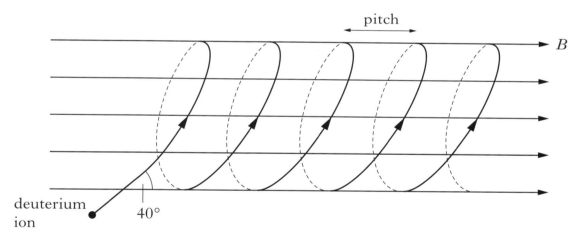

Figure 8C

   (i) Explain why the period of rotation for this deuterium ion is the same as in (*a*)(iii).   1

   (ii) The distance between adjacent loops in the helix is called the pitch.

   Calculate the pitch of the helical motion.   2

      (8)

[Turn over

Marks

9. Cyclotrons consist of two D shaped regions, known as Dees, separated by a small gap. An electric field between the Dees accelerates the charged particles. A magnetic field in the Dees causes the particles to follow a circular path. Negatively charged hydrogen ions (H⁻) are released from point A and follow the path as shown in Figure 9A.

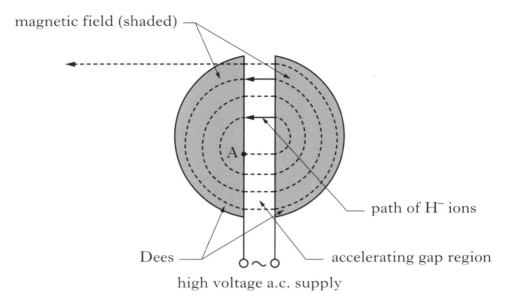

Figure 9A

(a) (i) State the direction of the magnetic induction B. **1**

   (ii) Explain why an a.c. supply must be used to provide the electric field. **1**

(b) The energy gained by the H⁻ ion for one transit of the gap region is 1.5×10^{-14} J.

   (i) How many transits of the gap would occur before relativistic effects should be taken into account? **2**

   (ii) When relativistic effects are taken into account, the magnitude of the magnetic induction must increase as the radius of the particle path increases.

   Explain why this is the case. **2**

Marks

9. (continued)

(*c*) The diagram shown in Figure 9B represents the paths of particles A, B and C in a perpendicular magnetic field.

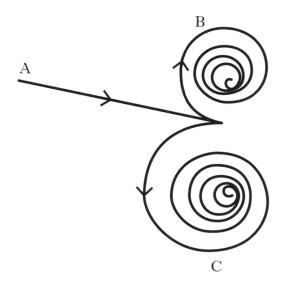

Figure 9B

(i) What information can be deduced about the particles A, B and C?   **2**

(ii) Why do the radii of paths B and C decrease?   **1**

(9)

[Turn over

Marks

10. An inductor of inductance 4·0 H with negligible resistance is connected in series with a 48 Ω resistor shown in Figure 10A.

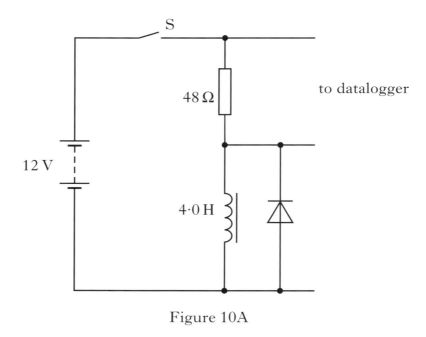

Figure 10A

The datalogger is set to display a graph of current against time.

(a) Sketch the graph obtained from the time the switch S is closed until the current reaches a maximum. Numerical values are required on the current axis only.

2

(b) Calculate the initial rate of change of current in the 4·0 H inductor.

2

(c) The 4·0 H inductor is now connected in the circuit shown in Figure 10B.

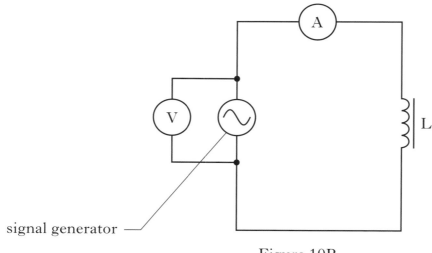

Figure 10B

The frequency of the signal generator is varied and the potential difference across the inductor is kept constant.

Explain what happens to the current in the inductor as the frequency of the signal generator is increased.

1

(5)

Marks

11. (*a*) When sunlight hits a thin film of oil floating on the surface of water, a complex pattern of coloured fringes is observed.

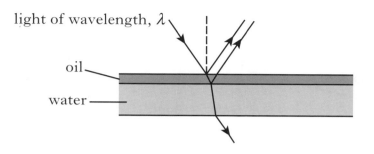

Figure 11

Explain how these fringes are formed.   2

(*b*) The surface of a lens is coated with a thin film of magnesium fluoride.

Calculate the minimum thickness required to make the lens non-reflecting at a wavelength of 555 nm.   2

(*c*) The lens of a digital camera appears to be purple in white light.

Explain this observation.   2

(6)

[Turn over

Marks

12. A series of coloured LEDs are used in the Young's slit experiment as shown in Figure 12. The distance from the slits to the screen is $(2{\cdot}50 \pm 0{\cdot}05)$m. The slit separation is $(3{\cdot}0 \pm 0{\cdot}1) \times 10^{-4}$ m.

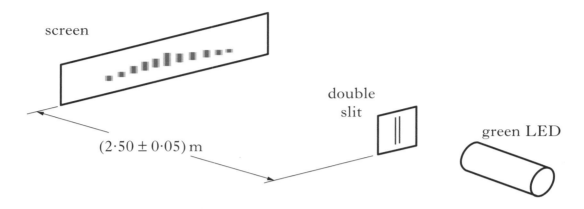

Figure 12

Colour of LED	Wavelength (nm)
Red	650 ± 2
Green	510 ± 2
Blue	470 ± 2

(a) State whether the pattern on the screen is caused by the division of wavefront or the division of amplitude.   **1**

(b) (i) Calculate the fringe separation observed on the screen when the green LED is used.   **2**

     (ii) Calculate the absolute uncertainty in the fringe separation.   **3**

     (iii) Which measurement has the most significant impact on the absolute uncertainty?

        Justify your answer.   **1**

                                                       (7)

Marks

13. A student, wearing polarising sunglasses, is using a tablet computer outdoors. The orientation of the tablet seems to affect the image observed by the student.

   Two orientations are shown in Figure 13A.

Landscape mode   Portrait mode

Figure 13A

(a) In landscape mode the image appears bright and in portrait mode it appears dark.

   (i) What may be concluded about the light emitted from the tablet screen? **1**

   (ii) The student slowly rotates the tablet. Describe the change in brightness observed by the student as it is rotated through 180°. **2**

(b) Unpolarised sunlight is incident on a water surface as shown in Figure 13B.

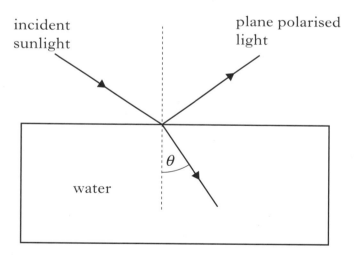

Figure 13B

   The light is 100% plane polarised on reflection.

   Calculate the angle of refraction θ. **2**

   (5)

[END OF QUESTION PAPER]

Acknowledgements

Permission has been sought from all relevant copyright holders and Hodder Gibson is grateful for the use of the following:
Image © aceshot1/Shutterstock.com (2013 paper page 4);
Image © ANDRITZ HYDRO GmbH (2014 page 3).

ADVANCED HIGHER | ANSWER SECTION

SQA ADVANCED HIGHER PHYSICS 2010–2014

1. (a) $v = \dfrac{s}{t}$ **or** $v = \dfrac{s}{t}$

$v = \dfrac{2\pi r}{T}$ $v = \dfrac{r\theta}{t}$

$\omega = \dfrac{2\pi}{T}$ $\omega = \dfrac{\theta}{t}$

$v = r\omega$ $v = r\omega$

(b) (i) $\omega = \dfrac{\theta}{t}$

$= \dfrac{3\cdot1}{4\cdot5} = 0\cdot69$

$v = r\omega$
$= 0\cdot69 \times 0\cdot148$
$= 0\cdot10 \text{ m s}^{-1}$

(ii) $\%\Delta\theta = \dfrac{0\cdot1}{3\cdot1} \times 100 = 3\cdot2\%$

$\%\Delta t = \dfrac{0\cdot1}{4\cdot5} \times 100 = 2\cdot2\%$

$\%\Delta r = \dfrac{0\cdot001}{0\cdot148} \times 100 = 0\cdot68\%$

$\%\Delta\omega = \sqrt{(\%\Delta\theta^2 + \%\Delta t^2 + \%\Delta r^2)}$
$= \sqrt{(3\cdot2^2 + 2\cdot2^2 + 0\cdot68^2)}$
$= 3\cdot9 \ (\%)$

(iii) (A) velocity changing
or
changing direction
(B) towards centre (of turntable)

2. (a) $F = mr\omega^2$
$= 0\cdot2 \times 0\cdot35 \times 6\cdot0^2$
$= 2\cdot5 \text{ N}$

(b) $\tan\theta = \dfrac{m\omega^2 r}{mg}$ or $= \dfrac{\omega^2 r}{g}$ or $= \dfrac{v^2}{rg}$

$= \dfrac{2\cdot5}{0\cdot2 \times 9\cdot8}$ **or** $\dfrac{6\cdot0^2 \times 0\cdot35}{9\cdot8}$ **or** $\dfrac{2\cdot1^2}{0\cdot35 \times 9\cdot8}$

$\theta = 52°$

(c) θ decreases
Centripetal force or $r\omega^2$ or v^2/r decreases

3. (a) $\alpha = \dfrac{\omega - \omega_0}{t}$

$= \dfrac{35 - 0}{0\cdot55}$

$= 64 \text{ rad s}^{-2}$

(b) θ = area under graph **or** $a_t = r\alpha$
$= \frac{1}{2} \times b \times h$ $= 0\cdot14 \times 64$
$= \frac{1}{2} \times 0\cdot55 \times 35$ $= 8\cdot96 \text{ ms}^{-2}$
$= 9\cdot6 \ (\text{rad})$ $s = ut + \dfrac{1}{2}at^2$
$s = r \times \theta$
$= 0\cdot14 \times 9\cdot6$ $= 0 + \dfrac{1}{2} \times 8\cdot96 \times 0\cdot55^2$
$= (1\cdot3 \text{ m})$ $= 1\cdot36 \text{ m (accept)}$
    rounding allowance

(c) $mgh = 2\cdot5 \times 9\cdot8 \times 1\cdot3$
$v = \omega \times r$
$= 35 \times 0\cdot14$
$= 4\cdot9 \ (\text{ms}^{-1})$

$mgh = \frac{1}{2}mv^2 + \frac{1}{2}I\omega^2$
$2\cdot5 \times 9\cdot8 \times 1\cdot3 = \frac{1}{2} \times 2\cdot5 \times 4\cdot9^2 + \frac{1}{2} \times I \times 35^2$
$31\cdot85 = 30\cdot01 + 612\cdot5 \times I$
$I = \dfrac{1\cdot84}{612\cdot5}$
$= 3\cdot0 \times 10^{-3} \text{ kg m}^2$

4. (a) **Total angular** momentum before (an event) = **total angular** momentum after (an event)
in the absence of external torques

(b) (i) $L = I \times \omega$
$= 4\cdot1 \times 2\cdot7$
$= 11 \text{ kg m}^2 \text{ s}^{-1}$ or $\text{kg m}^2 \text{ rad s}^{-1}$

(ii) $I_m = mr^2 = (2\cdot5 \times 0\cdot60^2) = \mathbf{0\cdot90} \text{ kg m}^2$
$I_T = (4\cdot1 + 0\cdot90)$ $= \mathbf{5\cdot0} \text{ kg m}^2$
$I_0 \times \omega_0 = I_T \times \omega_T$
$11 = 5\cdot0 \times \omega$
$\omega = 2\cdot2 \text{ rad s}^{-1}$

(iii) ω increased
r reduced / I reduced
since L or $I\omega$ constant

(c) $\alpha = \dfrac{\omega - \omega_0}{t}$

$= \dfrac{0 - 1\cdot5}{0\cdot75}$

$= -2\cdot0 \text{ rad s}^{-2}$

$\tau = I \times \alpha$
$= 4\cdot5 \times -2$
$= -9 \text{ N m}$

5. (a) $a = -\omega^2 y$
$-35 = -\omega^2 \times 0\cdot012$ **or** $35 = -\omega^2 \times (-0\cdot012)$
$\omega = 54 \text{ rad s}^{-1}$

(b) $y = 0\cdot012 \sin \text{ or } \cos 54t$

(c) $v = \dfrac{dy}{dt}$ **or** $\dfrac{d(0\cdot012 \sin 54t)}{dt}$

$v = (+) 0\cdot65 \cos 54t$
or
$v = -0\cdot65 \sin 54t$

(d) $E_k = \frac{1}{2}m\omega^2 A^2$
$= \frac{1}{2} \times 1\cdot4 \times 54^2 \times 0\cdot012^2$
$= 0\cdot29 \text{ J}$

6. (a) $E = \dfrac{Q}{4\pi\epsilon_0 r^2}$

$E = \dfrac{-1\cdot92 \times 10^{-12}}{4 \times \pi \times 8\cdot85 \times 10^{-12} \times (1\cdot00 \times 10^{-3})^2}$
$E = -1\cdot73 \times 10^4 \text{ NC}^{-1}$

(b) (i) Student A
(ii) E-field is zero inside a hollow conductor
E-field has inverse square dependence outside the conductor

(c) (i) $F = \dfrac{Q_1 Q_2}{4\pi\epsilon_o r^2}$

$F = \dfrac{(-)2\cdot97 \times 10^{-8} \times (-)1\cdot92 \times 10^{-12}}{4 \times \pi \times 8\cdot85 \times 10^{-12} \times (4\cdot12 \times 10^{-2})^2}$

$F = 3\cdot02 \times 10^{-7}$ N

(ii)

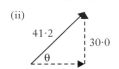

$\sin\theta = \dfrac{30\cdot0}{41\cdot2}$

$\theta = 46\cdot7°$

$F = F \times \sin\theta$
$F = 3\cdot02 \times 10^{-7} \times \sin 46\cdot7°$
$F = 2\cdot20 \times 10^{-7}$ N

(iii) Resultant F $= 4 \times F$
$= 8\cdot80 \times 10^{-7}$ N

(iv) Resultant F $= m \times g$
$8\cdot80 \times 10^{-7} = m \times 9\cdot8$
$m = 9\cdot0 \times 10^{-8}$ kg

7. (a) Magnetic field is out of the page

(b) $q \times v \times B = m \times \dfrac{v^2}{r}$

$r = \dfrac{m \times v}{q \times B}$

$= \dfrac{1\cdot673 \times 10^{-27} \times 6\cdot0 \times 10^6}{1\cdot6 \times 10^{-19} \times 0\cdot75}$

$r = 8\cdot4 \times 10^{-2}$ m

(c) $q \times E = q \times v \times B$
$E = v \times B$
$= 6\cdot0 \times 10^6 \times 0\cdot75$
$E = 4\cdot5 \times 10^6$ Vm^{-1}

(d) (i) $\frac{1}{2} \times m \times v^2 = \dfrac{qQ}{4\pi\epsilon_0 r}$

$r = \dfrac{qQ}{2 \times \pi \times \epsilon_0 \times m \times v^2}$

(ii) $r = \dfrac{qQ}{2 \times \pi \times \epsilon_0 \times m \times v^2}$

$= \dfrac{1\cdot60 \times 10^{-19} \times (29 \times 1\cdot6 \times 10^{-19})}{2 \times \pi \times 8\cdot85 \times 10^{-12} \times 1\cdot673 \times 10^{-27} \times (6\cdot0 \times 10^6)^2}$

$r = 2\cdot2 \times 10^{-13}$ m

(iii) Strong (nuclear) force.

(e) Reverse direction of magnetic field.
Reduce strength of magnetic field.

8. (a) $q \times V = \frac{1}{2} \times m \times v^2$
$1\cdot60 \times 10^{-19} \times 2400 = \frac{1}{2} \times 9\cdot11 \times 10^{-31} \times v^2$
$v^2 = 8\cdot43 \times 10^{14}$
$v = 2\cdot90 \times 10^7$ ms^{-1}

(b) (i) Constant force / acceleration / electric force / electric field in **vertical direction / upwards / downwards**
Constant motion / speed in horizontal direction

(ii) No unbalanced force / field / attraction acting on electron.

(c) (i) $E = \dfrac{V}{d}$

$= 100/0.01$
$= 10^4$ Vm^{-1}
$F = QE$
$= 1\cdot60 \times 10^{-19} \times 100/0\cdot01$
$= 1\cdot60 \times 10^{-15}$ N

$a = \dfrac{F}{m}$

$= \dfrac{1\cdot60 \times 10^{-15}}{9\cdot11 \times 10^{-31}}$

$= 1\cdot76 \times 10^{-15}$ ms^{-2}

(ii) $t = \dfrac{s_H}{v_H}$

$= 5\cdot17 \times 10^{-10}$ s

$v_v = u_v + a_v \times t$
$= 0 + 1\cdot76 \times 10^{15} \times 5\cdot17 \times 10^{-10}$
$v_v = 9\cdot10 \times 10^5$ ms^{-1}

(d) Length scanned decreases.
v_H increases / greater acceleration.
Shorter time between plates
or vertical speed is less on leaving the plates.

9. (a) $f = 2\cdot4$ Hz

(b) $\lambda = 0\cdot5$ m
$v = f\lambda$
$= 2\cdot4 \times 0\cdot5$
$= 1\cdot2$ m s^{-1}

(c) $y = 4\cdot3 \times 10^{-2} \sin 2\pi (2\cdot4t + 2\cdot0x)$

10. (a) Fringes produced by **interference** of light reflected from **top** and **bottom** surfaces of the film.
Different thicknesses / wavelengths / positions / angles affect colours.

(b) For non-reflecting coating,

Optical P.D. $= \lambda/2$
for **destructive** interference
Optical P.D. = 2nd
$2nd = \lambda/2$

$d = \dfrac{\lambda}{4 \times n}$

(c) $2 \times n \times d = (1\frac{1}{2}) \times \lambda$
$2 \times 1\cdot3 \times d = 1\cdot5 \times 7\cdot80 \times 10^{-7}$
$d = 4\cdot50 \times 10^{-7}$ m

(d) (i) $E_k = \frac{1}{2} \times m \times v^2$
$4\cdot12 \times 10^{-21} = \frac{1}{2} \times 1\cdot43 \times 10^{-25} \times v^2$

$v = \sqrt{\dfrac{2 \times 4\cdot12 \times 10^{-21}}{1\cdot43 \times 10^{-25}}}$

$v = 240$ ms^{-1}

$p_{ru} = m \times v$
$= 1\cdot43 \times 10^{-25} \times 240$
$p_{ru} = 3\cdot43 \times 10^{-23}$ kgms^{-1}

(ii) $p_{ph} = \dfrac{h}{\lambda}$

$= \dfrac{6\cdot63 \times 10^{-34}}{7\cdot80 \times 10^{-7}}$

$p_{ph} = 8\cdot50 \times 10^{-28}$ kgms^{-1}

(iii) $N \times p_{ph} = p_{ru}$
$N \times 8\cdot50 \times 10^{-28} = 3\cdot43 \times 10^{-23}$
$N = 4\cdot04 \times 10^4$ photons

11. (a) unpolarised light
$\Rightarrow$ (Electric field vector) oscillates or vibrates in all **planes**
polarised
$\Rightarrow$ (Electric field vector) oscillates or vibrates in **one plane**

(b) (i) $\theta = 0°$
$\theta = 180°$

(ii) Measure I and θ for at least 5 different values of θ.
Plot a graph of I against $\cos^2\theta$.
Graph produced is a straight line through the origin.

ADVANCED HIGHER PHYSICS 2011

1. (a) (i)
$$E = mc^2$$
$$2.08 \times 10^{-10} = m \times (3.0 \times 10^8)^2$$
$$m = \frac{2.08 \times 10^{-10}}{9.0 \times 10^{16}}$$
$$m = 2.3 \times 10^{-27} \text{ kg}$$

(ii) $m = m_o \times \left(\dfrac{1}{\sqrt{1 - \dfrac{v^2}{c^2}}} \right)$

$$2.3 \times 10^{-27} = 1.673 \times 10^{-27} \times \left(\frac{1}{\sqrt{1 - \dfrac{v^2}{(3.0 \times 10^8)^2}}} \right)$$

$$v = 2.1 \times 10^8 \text{ m s}^{-1}$$

(b) (i)
$$E_k = \tfrac{1}{2} m v^2$$
$$3.15 \times 10^{-21} = 0.5 \times 1.675 \times 10^{-27} \times v^2$$
$$v^2 = 3.76 \times 10^6$$
$$v = 1.94 \times 10^3 \ (m\ s^{-1})$$
$$p = m v$$
$$= 1.675 \times 10^{-27} \times 1.94 \times 10^3$$
$$= 3.25 \times 10^{-24} \text{ kg m s}^{-1} - \text{given}$$

(ii)
$$p = \frac{h}{\lambda}$$
$$3.25 \times 10^{-24} = \frac{6.63 \times 10^{-34}}{\lambda}$$
$$\lambda = \frac{6.63 \times 10^{-34}}{3.25 \times 10^{-24}}$$
$$\lambda = 2.04 \times 10^{-10} \text{ m}$$

(c) (i) Strong (nuclear) (force)
(ii) 10^{-14} m
(iii) Quark

2. (a) (i) $I_{rod} = 1/3\ m\ l^2$
$$= 1/3 \times 0.040 \times 0.30^2$$
$$= 1.2 \times 10^{-3} \text{ kg m}^2$$

(ii) $I_{wheel} = (5 \times I_{rod}) + m_{(rim)}\ r^2$
$$= (5 \times 1.2 \times 10^{-3}) + (0.24 \times 0.30^2)$$
$$= 6 \times 10^{-3} + 0.0216$$
$$= 0.0276$$
$$= 0.028 \text{ kg m}^2 - \text{given}$$

(b) (i) $v = \omega r$
$$19.2 = \omega \times 0.30$$
$$\omega = \frac{19.2}{0.30}$$
$$\omega = 64 \text{ rad s}^{-1}$$

(ii) A $\omega = \omega_o + \alpha t$
$$0 = 64 + \alpha \times 6.7$$
$$\alpha = -\frac{64}{6.7}$$
$$\alpha = -9.6 \text{ rad s}^{-2}$$

B $\tau = I \times \alpha$
$$= 0.028 \times (-)\ 9.6$$
$$= (-)\ 0.27 \text{ Nm}$$

3. (a) $\omega = \dfrac{2\pi}{T}$
$$= \frac{2 \times 3.14}{5.6 \times 24 \times 60 \times 60}$$
$$= 1.3 \times 10^{-5} \text{ rad s}^{-1} - \text{given}$$

(b) $F_C = F_G$
$$M_2 \omega^2 r = \frac{GM_1 M_2}{r^2}$$
$$2.0 \times 10^{30} \times (1.3 \times 10^{-5})^2 \times 3.6 \times 10^{10}$$
$$= \frac{6.67 \times 10^{-11} \times 2.0 \times 10^{30} \times M_1}{(3.6 \times 10^{10})^2}$$
$$M_1 = 1.2 \times 10^{32} \text{ kg}$$

(c) (i) $E_P = -\dfrac{GM_1 M_2}{r^2}$
$$= -\frac{6.67 \times 10^{-11} \times 2.0 \times 10^{30} \times 1.2 \times 10^{32}}{3.6 \times 10^{10}}$$
$$= -4.4 \times 10^{41} \text{ J} - \text{given}$$

(ii) $v = r\omega$
$$= 3.6 \times 10^{10} \times 1.3 \times 10^{-5}$$
$$= 4.68 \times 10^5$$
$$E_k = \tfrac{1}{2} mv^2$$
$$= \tfrac{1}{2} \times 2.0 \times 10^{30} \times (4.68 \times 10^5)^2$$
$$= 2.2 \times 10^{41} \text{ J}$$

(iii) $E_{total} = E_K + E_P$
$$E_{total} = 2.2 \times 10^{41} + (-4.4 \times 10^{41})$$
$$= -2.2 \times 10^{41} \text{ J}$$

(d) Frequency increases or blue shift when star approaches.
Frequency decreases or red shift when star recedes.

4. (a) $y = A \sin \omega t$
$$\omega = \frac{2 \times \pi}{T}$$
$$= \frac{2 \times 3.14}{5.7}$$
$$= 1.1$$
$$y = 2.9 \sin 1.1 t$$

(b) $a = -\omega^2 y$
$$= -1.1^2 \times (\pm)\ 2.9$$
$$= (\pm)\ 3.5 \text{ m s}^{-2}$$

(c) F_{max} occurs at $y = \pm 2.9$ m

(d) $E_k = \tfrac{1}{2} m \omega^2 (A^2 - y^2)$
$$= \tfrac{1}{2} \times 4.0 \times 10^4 \times 1.1^2 \times (2.9)^2$$
$$= 2.0 \times 10^5 \text{ J}$$

(e) (i) Period unaffected
(ii) Amplitude is reduced

5. (a) (i) Bring a <u>negative</u> charged rod close to the balloon,
earth (touch) sphere,
remove earth,
remove rod.

   or

   Touch 2 balloons together, bring charged rod near one,
separate balloons **before removing rod**, identify
which balloon is positive.

   (ii) $E = \dfrac{Q}{4\pi\varepsilon_o r^2}$

   $= \dfrac{(+)120 \times 10^{-6}}{4\pi \times 8.85 \times 10^{-12} \times (0.35)^2}$

   $E = (+)\,8.8 \times 10^6\ NC^{-1}\ \text{or}\ Vm^{-1}$

   (ii) E

   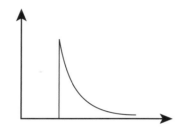

(b) (i) $F = qE$

   $E_w = Fd$

   $E_w = qV$

   $\cancel{q}V = \cancel{q}Ed$

   $V = Ed$

   $E = \dfrac{V}{d}$

   (ii) $V = E \times d$

   $V = 7.23 \times 10^4 \times 489$

   $V = 3.54 \times 10^7\ V$

   (iii) $I = \dfrac{Q}{t}\ \&\ P = IV$

   $I = \dfrac{5.0}{348 \times 10^{-6}}$

   $I = 14367.8\ A$

   $P = 14367.8 \times 3.54 \times 10^7$

   $P = 5.1 \times 10^{11}\ W$

(c)

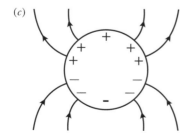

6. (a) (i) Increasing/changing current leads to increasing/changing
magnetic field causes a back emf.

   (ii) $E = -\dfrac{dI}{dt} L$

   $-12.0 = -\dfrac{dI}{dt} 0.6$

   $\dfrac{dI}{dt} = 20$

   $\dfrac{dI}{dt} = 20 As^{-1}$

   (iii) An inductor has an inductance of 1 Henry if an emf of
1 volt is induced when the current changes at a rate of
1 As^{-1}.

   (iv) Generates a **large** (back) emf

   due to **<u>rapid</u>** change
or **<u>collapse</u>** in **B-field**.

   (v) $V = IR$

   $12.0 = I \times 28$

   $I = \dfrac{12.0}{28}$

   $I = 0.43 A$

(b) $99 kmh^{-1} = \dfrac{99000}{3600} = 27.5 ms^{-1}$

   $v^2 = u^2 + 2as$

   $0^2 = 27.5^2 + 2 \times -1.0 \times s$

   $0 = 756.25 - 2s$

   $s = \dfrac{756.25}{2} = 378 m$

Yes before the signal.

(c) (i) Wavelength, $\lambda = \dfrac{v}{f_s}$

   $\lambda_{obs} = \dfrac{v}{f_s} - \dfrac{v_s}{f_s}$

   The observed frequency, $f_{obs} = \dfrac{v}{\lambda_{obs}} = \dfrac{v}{\frac{1}{f_s}(v - v_s)}$

   (ii) A $f_{obs} = f_s\left(\dfrac{v}{v - v_s}\right)$

   $f_{obs} = 294\left(\dfrac{340}{340 - 28.0}\right)$

   $f_{obs} = 320\ Hz$

   B $f_{obs} = f_s\left(\dfrac{v}{v + v_s}\right)$

   $f_{obs} = 294\left(\dfrac{340}{340 + 28.0}\right)$

   $f_{obs} = 272\ Hz$

7. (a) (i) Towards Y.

   Cancellation of B-field between the wires
or opposite magnetic fields caused by each wire cause
attraction.

(ii) $\dfrac{F}{L} = \dfrac{\mu_0 I_1 I_2}{2\pi r}$

$\dfrac{F}{L} = \dfrac{4\pi \times 10^{-7} \times 4.7 \times 4.7}{2\pi r \times 360 \times 10^{-3}}$

$\dfrac{F}{L} = 1.2 \times 10^{-5} \, Nm^{-1}$

(b) (i) $F = \dfrac{0.0058 + 0.0061 + 0.0063 + 0.0057 + 0.0058 + 0.0062}{6}$

$F = 0.0060 N$

$F = B1l$

$6.0 \times 10^{-3} = B \times 1.98 \times 0.054$

$B = \dfrac{6.0 \times 10^{-3}}{1.98 \times 0.054}$

$B = 0.056 T$

(ii) Scale Reading uncertainty (SRU)

± 1 digit $\Rightarrow \pm 0.0001 N$

Random uncertainty (RU)

$= \left(\dfrac{max - min}{n} \right)$

$= \left(\dfrac{0.0063 - 0.0057}{6} \right) = 0.0001 N$

$\Delta F = \sqrt{SRU^2 + RU^2 + calibration \ uncert^2}$

$\Delta F = \sqrt{0.0001^2 + 0.0001^2 + 0.00005^2}$

$= 1.5 \times 10^{-4} \, N$

(iii) $\dfrac{\Delta B}{B} = \sqrt{\left(\dfrac{\Delta F}{F}\right)^2 + \left(\dfrac{\Delta I}{I}\right)^2 + \left(\dfrac{\Delta l}{l}\right)^2}$

$\dfrac{\Delta B}{B} = \sqrt{\left(\dfrac{1.5 \times 10^{-4}}{0.0060}\right)^2 + \left(\dfrac{0.02}{1.98}\right)^2 + \left(\dfrac{0.0005}{0.054}\right)^2}$

$\dfrac{\Delta B}{B} = \sqrt{8.12 \times 10^{-4}}$

$\dfrac{\Delta B}{B} = 0.029$

$\Delta B = \pm 0.002 T$

8. (a) $F = BIl \, (\sin\theta)$ **or** $F = BIl$

but $I = \dfrac{q}{t}$

$v = \dfrac{l}{t}$

substituting to get

$F = B \dfrac{q}{t} vt$

(b) $F = \dfrac{mv^2}{r} = Bqv$

$v = \dfrac{Bqr}{m}$

$v = \dfrac{3.6 \times 10^{-3} \times 1.6 \times 10^{-19} \times 2.8 \times 10^{-3}}{9.11 \times 10^{-31}}$

$v = \dfrac{1.6128 \times 10^{-31}}{9.11 \times 10^{-31}}$

$v = 1.77 \times 10^6$

$v = v_{total} \times \sin\theta$

$\dfrac{1.77 \times 10^6}{2.0 \times 10^6} = \sin\theta$

$\theta = 62°$

(c) Radius decreases.
Pitch increases.

9. (a) Slits/gaps in horizontal and vertical direction.
Explanation of interference pattern.

(b) $\lambda = \dfrac{d\Delta x}{D}$

$4.88 \times 10^{-7} = \dfrac{d \times 8.0 \times 10^{-3}}{3.6}$

$d = 2.2 \times 10^{-4} m$

(c) (i) B

Larger λ gives larger Δx.

(ii) D

As horizontal d increases horizontal Δx decreases.

As vertical d decreases vertical Δx increases.

10. (a) A stationary wave is formed by the **interference** between waves, travelling in **opposite** directions or **reflecting** from the end supports.

(b) (i) $T = mg = 4.02 \times 9.8 = 39 N$

$f = \dfrac{1}{2l}\sqrt{\dfrac{T}{\mu}}$

$f = \dfrac{1}{2 \times 0.780}\sqrt{\dfrac{39}{1.92 \times 10^{-4}}}$

$f = 290 Hz$

Note is D.

(ii) $2 \times$ answer to **10.** (b) (i)
$f = 2 \times 290 = 580 \, Hz$

ADVANCED HIGHER PHYSICS
2012

1. (a) (i) $m = \dfrac{m_0}{\sqrt{1 - \dfrac{v^2}{c^2}}}$

$1.8 = \dfrac{1}{\sqrt{1 - \dfrac{v^2}{(3.0 \times 10^8)^2}}}$

$v = 2.5 \times 10^8 \text{ ms}^{-1}$

(ii) $E = mc^2$

$E = 1.8 \times 9.11 \times 10^{-31} \times (3.0 \times 10^8)^2$

$E = 1.5 \times 10^{-13}$ J

(iii) Weak force

(b) (i) Electron Diffraction.
Interference.
Fire electrons through crystals.
Thomson-Reid Experiment.

(ii) Compton Effect.
Photoelectric effect.
e/m experiment.
Electrons deflected in a deflection tube.
Electron's back scattering.

2. (a) (i) $a = \dfrac{dv}{dt}$

$\int dv = \int a.dt$

$v = at + c$

at $t = 0, c = u$

(ii) $v = u + at$

$v^2 = (u + at)(u + at)$

$v^2 = u^2 + 2uat + a^2t^2$

$v^2 = u^2 + 2a(ut + \dfrac{1}{2}at^2)$

$v^2 = u^2 + 2as$

(b) (i) $s = \dfrac{1}{2} \times 29.8 = 14.9$m

$v^2 = u^2 + 2as$

$9.64^2 = 0^2 + 2 \times a \times 14.9$

$a = 3.12 \text{ ms}^{-2}$

(ii) $v^2 = u^2 + 2as$
$10.9^2 = 9.64^2 + 2a(\pi \times 8.20)$
$a = 0.5 \text{ ms}^{-2}$
$a = r\alpha$
$0.5 = 8.2 \times \alpha$
$\alpha = 0.061 \text{ rad s}^{-2}$

(c) (i) An indication of the central/ inward force
Provided by horizontal **component** of F.

(ii) Central force is no longer large enough to maintain her circular motion.

3. (a) (i) $y = A \sin \omega t$ **or** $y = A \cos \omega t$
$\omega = 2\pi f$
$\omega = 2\pi \times 0.76$
$\omega = 4.8 \text{ rad s}^{-1}$
$A = 0.18$ m
$A = 0.18 \sin 4.8t$

(ii) $g = \pm A\omega^2$
$9.8 = \pm A \times 4.8^2$
$A = \pm 0.43$ m

(b) (i) Assume diver 2 rods about one end
33.0 kg per rod

$I = \dfrac{1}{3} \text{ ml}^2$

$I = \dfrac{1}{3} \times 33.0 \times 0.90^2 = 8.9 \text{ kg m}^2$

$I = \dfrac{1}{3} \times 33.0 \times 0.94^2 = 9.7 \text{ kg m}^2$

$I = 18.6 \text{ kg m}^2$

(ii) Some indication of uneven mass distribution.

(iii) $L = I_1\omega_1 = I_2\omega_2$
$10.25 \times 0.55 = 7.65 \times \omega_2$
$\omega_2 = 0.74 \text{ rad s}^{-1}$

(c) (i) $E_{krot} = \dfrac{1}{2} I_1\omega_1^2$

$E_{krot} = \dfrac{1}{2} I_2\omega_2^2$

$\dfrac{1}{2} \times 10.25 \times 0.55^2 = 1.55$ J

$\dfrac{1}{2} \times 7.65 \times 0.74^2 = 2.09$ J

$\Delta E_{krot} = 0.54$ J

(ii) Work is being done by the diver.

4. (a) $mg = \dfrac{GM_p m}{r^2}$

(b) (i) $g = \dfrac{GM}{r^2}$

from graph $r = 1.2 \times 10^6$ m

$g = \dfrac{6.67 \times 10^{-11} \times 1.27 \times 10^{22}}{(1.2 \times 10^6)^2}$

$g = 0.59 \text{ N kg}^{-1}$ **or** ms^{-2}

(ii) $E = -\dfrac{GMm}{r}$

$E = -\dfrac{6.67 \times 10^{-11} \times 1.27 \times 10^{22} \times 112}{1.80 \times 10^6}$

$E = -5.27 \times 10^7$ J

(c) $\dfrac{GM_p m}{x^2} = \dfrac{GM_c m}{(1.96 \times 10^7 - x)^2}$

$\dfrac{7}{1} = \dfrac{M_p}{M_c}$ or $M_C = 1.81 \times 10^{21}$

$\dfrac{7}{1} = \dfrac{x^2}{(1.96 \times 10^7 - x)^2}$

$x = 1.42 \times 10^7$ m from Pluto

5. (a) Electrons/negative charges in sphere move to rhs of sphere leaving +ve charge on lhs of sphere.

(b) (i) $V = 3.0 \times 10^3 - (-2.0 \times 10^3)$ V
$= 5.0 \times 10^3$ V
$E = V / d$
$= 5.0 \times 10^3 / 0.042$
$= 1.2 \times 10^5$ $V m^{-1}$ or NC^{-1}

(ii) $F = qE$

$\quad = 140 \times 10^{-9} \times 1.2 \times 10^{5}$

$\quad = 1.7 \times 10^{-2}$ N

(iii) $\tan\theta = \dfrac{F}{mg} = \dfrac{1.7 \times 10^{-2}}{3.92 \times 10^{-2}}$

$\quad \theta = 23.4°$

$T = \dfrac{F}{\sin\theta} = \dfrac{1.7 \times 10^{-2}}{\sin 23.4}$

or $T = \dfrac{mg}{\cos\theta}$ as an alternative

$T = 4.3 \times 10^{-2}$ N

or

$T^{2} = (1.7 \times 10^{-2})^{2} + (3.92 \times 10^{-2})^{2}$
$T = 4.3 \times 10^{-2}$ N

$\tan\theta = \dfrac{1.7 \times 10^{-2}}{3.92 \times 10^{-2}}$

$\theta = 23.4°$

(c) Angle is unchanged.
Uniform electric field/force acting is constant.

6. (a) (i) Point Q

(ii) $Bqv = \dfrac{mv^{2}}{r}$

$\dfrac{q}{m} = \dfrac{v}{rB}$

(iii) $q / m = \dfrac{v}{rB}$

$\quad = \dfrac{2.29 \times 10^{6}}{0.0190 \times 2.50}$

$\quad = 4.82 \times 10^{7}$ C kg^{-1}

Alpha particle $q / m = \dfrac{3.20 \times 10^{-19}}{6.645 \times 10^{-27}}$

$\quad = 4.82 \times 10^{7}$ C kg^{-1}
Particle is alpha

(iv) $t = d / v$

$\quad = \pi r / v = \dfrac{3.14 \times 0.019}{2.29 \times 10^{6}}$

$\quad = 2.61 \times 10^{-8}$ s

(v) t is constant.
Both v and **r double** or are **directly proportional**

(b) (i) Particle is negatively charged.

(ii) Charge to mass ratio is smaller.

7. (a) (i) $\lambda = \dfrac{h}{p}$

$\quad = \dfrac{6.63 \times 10^{-34}}{6.26 \times 10^{-29}}$

$\quad = 1.06 \times 10^{-5}$ m

(ii) If an e.m.f. of ± 0.1 V is induced when the current is changing at the rate of 1 A s^{-1}, the inductance is 0.1 H.

(b) (i) Magnetic field strength increases and reaches a maximum value/levels off.

(ii) At $t = 0, dI / dt$ 4.0 A s^{-1}. At this time $E = -12$ V
$\quad E = -L\, dI / dt$
$\quad -12 = -L \times 4$
$\quad L = 3.0$ H

(iii)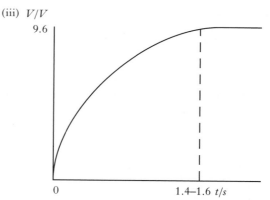

Maximum current $= E / R_{t} = 12 / 10 = 1.2$ A
Maximum p.d. across 8Ω
$\quad = 12 \times 8 = 9.6$ V

(iv) $E = \dfrac{1}{2} LI^{2}$

$\quad = 0.5 \times 3 \times 1.2^{2}$
$\quad = 2.2$ J

(c) Reading on A_{1} will increase as capacitive reactance decreases/$I \propto f$

Reading on A_{2} will decrease as inductive reactance increases/$I \propto 1/f$

8. (a) $B = \dfrac{\mu_{0}I}{2\pi r}$

$\quad = \dfrac{4 \times \pi \times 10^{-7} \times 25}{2 \times 3.14 \times 0.006}$

$\quad = 8.3 \times 10^{-4}$ T

(b) (i) The current in each wire produces a magnetic field. Same direction of the magnetic fields between the wires.

(ii) $\dfrac{F}{1} = \dfrac{\mu_{0}I_{1}I_{2}}{2\pi r}$

$\quad = \dfrac{4 \times \pi \times 10^{-7} \times 25 \times I}{2\pi r}$

$\quad = \dfrac{5.0 \times 10^{-6}\, I}{r}$

(iii) Weight per unit length $= 5.7 \times 10^{-3} \times 9.8$
$\quad = 0.056$ N m^{-1}

$\quad \dfrac{5.0 \times 10^{-6}\, I}{0.006} = 0.056$

$\quad I = 67$ A

9. (a) (i) Read from the graph $\lambda = 0.25$ m

(ii) Read from the graph $A = 0.58$ m

(iii) $v = f\lambda$
$\quad 1.25 = f \times 0.25$
$\quad f = 5.0$ Hz

(iv) $\Delta = \dfrac{2\pi x}{\lambda}$

$\quad \Delta = \dfrac{2\pi \times (0.44 - 0.25)}{0.25}$

Phase angle $= 1.5\pi = 4.7$ rad

(b) $y = (\pm)0.58\sin 2\pi(5.0t - \dfrac{x}{0.25})$

(c) $y = (\pm)0.29\sin(31t + 25x)$

10. (a) (i) $v = ds / dt$
$= 8.2t$
$= 8.2 \times 2$
$= 16 \text{ m s}^{-1}$

(ii) Frequency is increasing/increases.

Waves become <u>more and more</u> squashed together as speed increases or time between wave continually decreasing.

(b) $f = \dfrac{v}{v - v_s} \times f_s$

$= \dfrac{340}{340 - 16} \times 595$

$= 624 \text{ Hz}$

11. (a) (i) $\Delta x = \lambda D / d$

Gradient of graph

$= \dfrac{(1.30 - 0.30)}{(23 - 6) \times 10^{-3}}$

$= 58.8$

$d = 529 \times 10^{-9} \times 58.8$

$d = 3.1 \times 10^{-5} \text{ m}$

(ii) Uncertainty too small.

(iii) *Any two from:*
- Measure distance between several spots
- Use a bigger range of values for D
- Increase value(s) of D
- Additional data points.

(b) Spots blurred/elongated in horizontal direction. Spacing increased.

ADVANCED HIGHER PHYSICS 2013

1. (a) $\dfrac{mv^2}{r} = mg$

$\dfrac{v^2}{r} = g$

$9.8 = \dfrac{v^2}{6.2}$

$v = 7 \cdot 8 \text{ ms}^{-1}$

(b) (i)

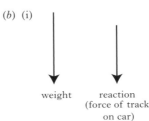

weight   reaction (force of track on car)

(ii) $\dfrac{mv^2}{r} = 11000 \text{ N}$

Weight $= mg = 870 \times 9.8$
$= 8500 \text{N}$

$R = 11000 - 8500$
$= 2500 \text{N}$

(c) By differentiation
$v = 9.1 - 6 \cdot 4t$
for $v = 0$, $t = 1 \cdot 4$ s
Max displacement,
$s = 9 \cdot 1t - 3.2t^2$
$s = (9.1 \times 1.4) - (3.2 \times 1.4^2)$
$s = 6.5$ m

2. (a) (i) Unbalanced torque $= I\alpha$
$= 54 \times 2.4$
$= 129.6 \text{ Nm}$

Applied torque $= 129.6 + 25$
$= 154.6 \text{ Nm}$

Applied torque $= F \times r$
$154.6 = F \times 1.2$
$F = 130 \text{ N}$

(ii) $\alpha = \dfrac{T}{I} = \dfrac{-25}{54} = -0.46 \text{ rads}^{-2}$

$\omega = \omega_0 + \alpha t$
$0 = \omega_0 + (-0.46 \times 3.6)$
$\omega_0 = 1.67 \text{ rads}^{-1}$
$\theta = \omega_0 t + \frac{1}{2}\alpha t^2$
$= (1.67 \times 3.6) + (0.5 \times -0.46 \times 3.6^2)$
$= 3.0 \text{ rad}$

2. (b) Acceleration is less

Applied torque is less
or
Applied force perpendicular to door is less

3. (a) (i) $\dfrac{mv^2}{r} = \dfrac{GMm}{r^2}$ **or** $mv^2 = \dfrac{GMm}{r^2}$

$\dfrac{v^2}{r} = \dfrac{GM}{r^2}$ $\left(rw^2 = \dfrac{GMm}{r^2}\right)$

subs for $w = \dfrac{2\pi}{T}$

$v = \dfrac{2\pi r}{T}$

$r = \dfrac{G\,m}{v^2} = \dfrac{G\,M\,T^2}{4\pi^2 r^2}$

$r^3 = \dfrac{G\,M\,T^2}{4\pi^2}$

(ii) $r^3 = GM\dfrac{T^2}{4\pi^2}$

$= \dfrac{6.67 \times 10^{-11} \times 1.7 \times 10^{30} \times (14 \times 24 \times 3600)^2}{4 \times \pi^2}$

$r = 1.6 \times 10^{10}\,\text{m}$

(b) $mg = \dfrac{G\,Mm}{r^2}$

$g = \dfrac{G\,M}{r^2} = \dfrac{6.67 \times 10^{-11} \times 5.4 \times 10^{26}}{(1.2 \times 10^8)^2}$

$= 2.5\,\text{N kg}^{-1}$

(c) (i) An object with an escape velocity greater than the speed of light.

(ii) escape velocity $v = \sqrt{\dfrac{2GM}{r}}$

$= \sqrt{\dfrac{2 \times 4.2 \times 10^{30} \times 6.67 \times 10^{-11}}{2.6 \times 10^4}}$

$= 1.5 \times 10^8\,\text{m s}^{-1}$

not a black hole

4. (a) Acceleration/unbalanced force is directly proportional to displacement
And in the opposite direction/directed towards the equilibrium position.

(b) (i) a = 1.28 m s⁻² (from graph)
T = 3.0s
$a = (-)\omega^2 y$

$\omega = \dfrac{2\pi}{T}$

$= 2.1\,(\text{rad s}^{-1})$
$1.28 = (-)2.1^2 y$
$y = 0.29\,\text{m}$

(ii) $v = (\pm)\omega\sqrt{A^2 - y^2}$

$= (\pm)2.1\sqrt{0.29^2 - 0.10^2}$

$= (\pm)\,0.57\,\text{m s}^{-1}$

(c) $E_k = E_p$
$\tfrac{1}{2}m\omega^2 A^2 - \tfrac{1}{2}m\omega^2 y^2 = \tfrac{1}{2}m\omega^2 y^2$
$\tfrac{1}{2}m\omega^2 A^2 = m\omega^2 y^2$
$y^2 = 0.5 \times 0.29^2$
$y = 0.21\,\text{m}$

5. (a) (i) $mvr = \dfrac{nh}{2\pi}$

$9.11 \times 10^{-31} \times v \times 4.8 \times 10^{-10} = \dfrac{3 \times 6.63 \times 10^{-34}}{2 \times \pi}$

$v = 7.2 \times 10^5\,\text{m s}^{-1}$

(ii) $\lambda = \dfrac{h}{p}$

$= \dfrac{h}{mv}$

$= \dfrac{6.63 \times 10^{-34}}{9.11 \times 10^{-31} \times 7.2 \times 10^5}$

$= 1.0 \times 10^{-9}\,\text{m}$

(iii) Quantum mechanics

(b) $F = \dfrac{Q_1 Q_2}{4\pi\epsilon_0 r^2}$

$= \dfrac{(1.6 \times 10^{-19})^2}{4 \times \pi \times 8.85 \times 10^{-12} \times (5.3 \times 10^{-11})^2}$

$= 8.2 \times 10^{-8}\,(\text{N})$

$F = \dfrac{Gm_1 m}{r^2}$

$= \dfrac{6.67 \times 10^{-11} \times 9.11 \times 10^{-31} \times 1.673 \times 10^{-27}}{(5.3 \times 10^{-11})^2}$

$= 3.6 \times 10^{-47}\,(\text{N})$
ie electrostatic force is much greater than the gravitational force

6. (a) (i) $q \times V = \tfrac{1}{2} \times m \times v^2$
$1.60 \times 10^{-19} \times 4.0 \times 10^6 = \tfrac{1}{2} \times 1.673 \times 10^{-27} \times v^2$
$v^2 = 7.65 \times 10^{14}$
$v = 2.8 \times 10^7\,\text{ms}^{-1}$

(ii) $\tfrac{1}{2} \times m \times v^2 = \dfrac{qQ}{4\pi\epsilon_0 r}$

$r = \dfrac{qQ}{2 \times \pi \times \epsilon_0 \times m \times v^2}$

(iii) $Q = 40 \times 1.6 \times 10^{-19}$
$= 6.4 \times 10^{-18}\,\text{C}$

$r = \dfrac{qQ}{2 \times \pi \times \epsilon_0 \times m \times v^2}$

$= \dfrac{1.6 \times 10^{-19} \times 6.4 \times 10^{-18}}{2 \times \pi \times 8.85 \times 10^{-12} \times 1.673 \times 10^{-27} \times (2.8 \times 10^7)^2}$

$= 1.4 \times 10^{-14}\,\text{m}$

(b)

$$m = \frac{m_o}{\sqrt{1 - \frac{v^2}{c^2}}}$$

$$= \frac{1 \cdot 673 \times 10^{-27}}{\sqrt{1 - \left(\frac{0.8c}{c}\right)^2}}$$

$$= 2.8 \times 10^{-27} \text{ kg}$$

$$E = mc^2$$

$$= 2.8 \times 10^{-27} \times (3 \times 10^8)^2$$

$$= 2.5 \times 10^{-10} \text{ J}$$

(c) Spanning less than 10^{-14} m.
Strong force attractive.
Strong force much greater than electrostatic.

7. (a) One tesla is the magnetic induction of a magnetic field in which a conductor of length one metre carrying a current of one ampere is acted on by a force of one newton.

(b) $F = BIl$
$F = 0.20 \times 2.5 \times 0.4$
$F = 0.20$ N

(c) Flow rate will fall.
$F = BIl \sin \theta$ explanation.
Force will be reduced.

(d) $B = \frac{\mu_o I}{2\pi}$

$$B = \frac{4\pi \times 10^{-7} \times 0.6}{2 \times \pi \times 0.75}$$

$$B = 1.6 \times 10^{-7} \text{ T}$$

(e) (i) $\frac{F}{l} = \frac{\mu_0 \times I_1 \times I_2}{2\pi r}$

$$\frac{F}{l} = \frac{4\pi \times 10^{-7} \times 0.6 \times 0.6}{2\pi \times 0.010}$$

$$= 7.2 \times 10^{-6} \text{ N m}^{-1}$$

(ii) To the left/towards AB or a diagram.
Interaction between the magnetic fields.
Magnetic field lines between wires are in the opposite direction.

8. (a) Charge is quantised

(b)

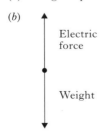

Electric force

Weight

(c) $E = \frac{V}{d}$

$$F = EQ = mg$$

$$m = \frac{QV}{gd}$$

$$= \frac{2.4 \times 10^{-18} \times 2000}{9.8 \times 0.016}$$

$$= 3.1 \times 10^{-14} \text{ kg}$$

9. (a) $V = \frac{Q}{4\pi\epsilon_0 r}$

$$V = \frac{-15 \times 10^{-6}}{4 \times 3.14 \times 8.85 \times 10^{-12} \times 0.65}$$

$$V = -2.1 \times 10^5 \text{ V}$$

(b) $\%\Delta r = \frac{0.02}{0.65} \times 100 = 3\%$

$$\%\Delta Q = \frac{0.4}{15} \times 100 = 2.7\%$$

$$\%\Delta V = (\pm)\sqrt{\%\Delta r^2 + \%\Delta Q^2}$$

$$= (\pm)\sqrt{9 + 7.1}$$

$$= (\pm)4.0\%$$

$$\Delta V = (\pm)\frac{4.0}{100} \times 2.1 \times 10^5 = (\pm)8 \times 10^3 \text{ V}$$

(c) $V = -2.1 \times 10^5$ V

10. (a) (i)

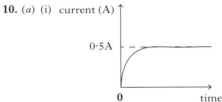

(ii) Changing magnetic field
Produces a back e.m.f in the inductor.

(b) $E = -L\frac{dI}{dt}$

$$E = -9.0 \text{ V}$$

$$\frac{dI}{dt} = \frac{E}{-L} = \frac{-9.0}{-0.40}$$

$$\frac{dI}{dt} = 23 \text{ As}^{-1}$$

11. (a) $\Delta x = \frac{1.2 \times 10^{-3}}{5} = 2.4 \times 10^{-4}$

$$\Delta x = \frac{\lambda L}{2d}$$

$$2.4 \times 10^{-4} = \frac{\lambda \times 0.05}{2 \times 6.2 \times 10^{-5}}$$

$$\lambda = 6.0 \times 10^{-7} \text{ m}$$

(b)

Spacing of fringes decreases from left to right.
or
Width of fringes decreases from left to right.

(c) (i)

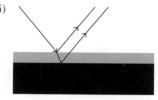

The two reflected rays interfere destructively

(ii) $d = \dfrac{\lambda}{4n}$

$= \dfrac{589 \times 10^{-9}}{4 \times 1.38}$

$= 1.07 \times 10^{-7}$ m

12. (a) $\varphi = \dfrac{2\pi x}{\lambda}$

$3.5 = \dfrac{2\pi \times 0.28}{\lambda}$

$\lambda = 0.50$ m

(b) (i) $\lambda = 0.25$ m

$v = f\lambda$

$= 5.0 \times 0.25$

$= 1.3$ m s^{-1}

(ii) $\dfrac{I_1}{A_1{}^2} = \dfrac{I_2}{A_2{}^2}$

$I \propto A^2$

$\dfrac{I_1}{0.03^2} = \dfrac{5I_1}{A_2{}^2}$

$A_2 = 0.07$ m

13. (a) In plane polarised light (the electric field vector of the light) vibrates/oscillates in one plane.

(b) Sound waves are not transverse waves.

(c) (i)

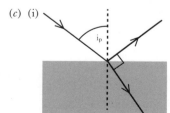

At the Brewster angle, the reflected light is plane polarised.

(ii) n = tan i$_p$
    = tan 54.5
    = 1.40

ADVANCED HIGHER PHYSICS 2014

1. (a) (i) $v = r\omega$

$8.8 = 7.8\,\omega$

$\omega = 1.1$ rad s^{-1}

(ii) $E_{rot} = \dfrac{1}{2} I\omega^2$

$100 \times 10^6 = \dfrac{1}{2} \times I \times 1.1^2$

$I = 1.7 \times 10^8$ kg m^2

(iii) There are no other moving parts in the blade system (e.g. gears) **OR** blades are rigid/do not bend.

(b) $\alpha = \dfrac{\omega - \omega_0}{t}$

$\alpha = \dfrac{1.1 - 0}{42}$

$= 0.026$

$T = I\alpha T$

$= 1.7 \times 10^8 \times 0.026$

$= 4.4 \times 10^6$ Nm

2. (a) $I = \dfrac{1}{2} mr^2$

$I = \dfrac{1}{2} \times 0.115 \times 0.015^2$

$I = 1.3 \times 10^{-5}$ kg m^2

(b) $\omega = \dfrac{v}{r}$

$\omega = \dfrac{1.60}{0.015}$

$\omega = 1.1 \times 10^2$ (rads^{-1})

$mgh = \dfrac{1}{2} mv^2 + \dfrac{1}{2} I\omega^2$

$0.28 = 0.1472 + \dfrac{1}{2} I(1.1 \times 10^2)^2$

$I = \dfrac{2 \times 0.1328}{(1.1 \times 10^2)^2}$

$I = 2.2 \times 10^{-5}$ kg m^2

(c) energy is lost
or
calculation assumes no energy is lost

3. (a) (i) The minimum velocity/speed that a mass must have to escape the gravitational field (of a planet).

(ii) $E_k + E_p = 0$

Therefore $\dfrac{1}{2} mv^2 - \dfrac{GMm}{r} = 0$

$v^2 = \dfrac{2GM}{r}$

$v = \sqrt{\dfrac{2GM}{r}}$

(iii)
$$v = \sqrt{\frac{2GM}{r}}$$

$$v = \sqrt{\frac{2 \times 6.67 \times 10^{-11} \times 6.0 \times 10^{24}}{1.7 \times 6.4 \times 10^6}}$$

$$= 8.6 \times 10^3 \text{ms}^{-1}$$

(b) $\dfrac{8.6 \times 10^3}{6} = 1.4 \times 10^3 \text{ ms}^{-1}$

Nitrogen, Oxygen, Methane, Carbon Dioxide could all be found on planet.

4. (a) The unbalanced force/ acceleration is proportional to the displacement of the object and acts in the opposite direction.

(b) 0.07 m

(c) (i) $\omega = \dfrac{\theta}{t}$

$$\omega = \frac{1500 \times 2\pi}{60}$$

$$\omega = 157 \text{ (rad s}^{-1})$$

$$a = (-) \omega^2 y$$

$$= (-) 157^2 \times 0.070$$

$$= (-) 1.7 \times 10^3 \text{ m s}^{-2}$$

(ii) $E_k = \dfrac{1}{2} m\omega^2 (A^2 - y^2)$

or

$$E_k = \frac{1}{2} m\omega^2 A^2$$

$$= \frac{1}{2} \times 1.40 \times 157^2 \times (0.070^2)$$

$$= 85 \text{ J}$$

5. (a) (i) Electrons behave like waves

(ii) Photoelectric effect or Compton scattering.
Collision and transfer of energy

(b) $\lambda = \dfrac{h}{p}$ or $\lambda = \dfrac{h}{mv}$

$$\lambda = \frac{6.63 \times 10^{-34}}{4.4 \times 10^6 \times 9.11 \times 10^{-31}}$$

$$\lambda = 1.7 \times 10^{-10} \text{ m}$$

(c) $\lambda = \dfrac{h}{p}$

$$\lambda = \frac{6.63 \times 10^{-34}}{300 \times 0.02}$$

$$\lambda = 1.1 \times 10^{-34} \text{ m}$$

This value is so small that no diffraction would be seen.
Or the de Broglie wavelength of the bullet is much smaller than the gap.

(d) (i) Electron orbits a nucleus / proton.

Angular momentum quantised
or
Certain allowed orbits / discrete energy levels.

(ii) $mvr = \dfrac{nh}{2\pi}$

$$= \frac{3 \times 6.63 \times 10^{-34}}{2 \times 3.14}$$

$$= 3.17 \times 10^{-34} \text{ kg m}^2 \text{ s}^{-1}$$

6. (a) (i) $F_{PA} = \dfrac{Q_1 Q_2}{4\pi\varepsilon_0 r^2}$

$$= \frac{(4.0 \times 10^{-9}) \times (2.6 \times 10^{-9})}{4\pi \times 8.85 \times 10^{-12} \times (2.5 \times 10^{-3})^2}$$

$$= 0.015 \text{ N}$$

$$(1.5 \times 10^{-2} \text{ N})$$

(ii)
$$(F_{RA}) = (-) 0.012 \text{ (N)}$$

$$(F_{SA}) = (-) 0.015 \text{ (N)}$$

F (due to P&S) $= 2 \times 0.015 \cos 37$
$$= 0.024 \text{N (to right)}$$

F (due to Q&R) $= 2 \times 0.012 \cos 37$
$$= 0.019 \text{N (to right)}$$

Combined force $= 0.024 + 0.019$
$$= 0.043 \text{ N to right}$$
$$(4.3 \times 10^{-2} \text{ N to right})$$

OR

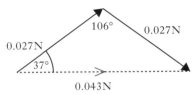

(b) (i) $V = \dfrac{Q}{4\pi\varepsilon_0 r}$

$$-37 = \frac{-2.6 \times 10^{-9}}{4\pi \times 8.85 \times 10^{-12} r}$$

$$r = 0.63 \text{ m}$$

(ii) The work done (energy) used would be the same
or
The charge is in a conservative field, the path taken between two points does not affect the work done (energy used).

7. (a) (i) $qV = \dfrac{1}{2} mv^2$

$$\frac{q}{m} = \frac{v^2}{2V}$$

(ii) **(A)**
$$\frac{q}{m} = \frac{v^2}{2V}$$

$$\frac{(2.92 \times 10^7)^2}{2 \times 2480} = 1.72 \times 10^{11} \text{ Ckg}^{-1}$$

$$\frac{(2.73 \times 10^7)^2}{2 \times 2150} = 1.73 \times 10^{11} \text{ Ckg}^{-1}$$

$$\frac{(2.61 \times 10^7)^2}{2 \times 2000} = 1.70 \times 10^{11} \text{ Ckg}^{-1}$$

$$\frac{(2.47 \times 10^7)^2}{2 \times 1750} = 1.74 \times 10^{11} \text{ Ckg}^{-1}$$

$$\frac{(2.26 \times 10^7)^2}{2 \times 1560} = 1.64 \times 10^{11} \text{ Ckg}^{-1}$$

(B)

$$\frac{1.72 + 1.73 + 1.70 + 1.74 + 1.64}{5} = 1.71 \times 10^{11} \text{ Ckg}^{-1}$$

(C)

Drawing a graph of v^2 vs $2V$
and calculate the gradient $= \dfrac{q}{m}$

7. (b) (i)
$$QV = \frac{1}{2}mv^2$$
$$1.60 \times 10^{-19} \times 2.08 \times 10^3 = \frac{1}{2} \times 9.11 \times 10^{-31} \, v^2$$
$$v = 2.70 \times 10^7 \text{ ms}^{-1}$$

(ii)
$$t = \frac{d}{v}$$
$$t = \frac{85.0 \times 10^{-3}}{2.70 \times 10^7}$$
$$t = 3.15 \times 10^{-9} \text{ (s)}$$
$$QE = ma$$
$$E = \frac{1900}{0.105} = 1.8 \times 10^4 \text{ (Vm}^{-1})$$
$$1.60 \times 10^{-19} \times \frac{1900}{0.105} = 9.11 \times 10^{-31} \, a$$
$$a = 3.18 \times 10^{-15} \text{ (m s}^{-2})$$
$$s = ut + \frac{1}{2}at^2$$
$$s = 0 + \frac{1}{2} \times 3.18 \times 10^{15} (3.15 \times 10^{-9})^2$$
$$s = 0.0157 \text{ m}$$

(iii) Lower than x /vertical displacement reduces

Increased horizontal velocity.

Time between plates reduced.

8. (a) (i) $qvB = \dfrac{mv^2}{r}$

$$r = \frac{mv}{qB}$$

(ii)
$$r = \frac{mv}{qB}$$
$$r = \frac{0.5}{2}$$
$$= 0.25 \text{ m}$$
$$0.25 = \frac{3.343 \times 10^{-27} \times 2.4 \times 10^7}{1.60 \times 10^{-19} \text{ x } B}$$
$$B = 2.0\text{T}$$

(iii)
$$T = \frac{2\pi r}{r}$$
$$T = \frac{2\pi \times 0.25}{2.4 \times 10^7}$$
$$T = 6.5 \times 10^{-8}\text{s}$$

(b) (i) *Any one from:*
- Period independent of velocity.
- Radius and vertical velocity reduce (in proportion) or both reduce
- Angular velocity is constant
- The magnetic induction has not changed

(ii)
$$v_h = v \cos\theta$$
$$v_h = 2.4 \times 10^7 \cos 40$$
$$v_h = 1.84 \times 10^7 \text{ (m s}^{-1})$$
$$pitch = v \times t$$
$$pitch = 1.84 \times 10^7 \times 6.5 \times 10^{-8}$$
$$pitch = 1.2 \text{ m}$$

9. (a) (i) Out of page

(ii) To ensure that the accelerating potential is in the correct direction for the particle's motion.

or

Direction of force acting on charges reversed.

(b) (i)
$$E_k = \frac{1}{2}mv^2$$
$$\frac{1}{2} \times 1.673 \times 10^{-27} \times (3.0 \times 10^7)^2$$
$$= 7.5 \times 10^{-13} \text{ J}$$
$$\textit{Number of gap transits} = \frac{7.5 \times 10^{-13}}{1.5 \times 10^{-14}} = 50$$

(ii) Their mass will increase (become relativistic)

A greater centripetal force will be required

or

To keep the radius of orbit within the dimensions of the cyclotron

$$\left(qvB = \frac{mv^2}{r} \right)$$

(c) (i) A has no charge.

B & C have different charge to mass ratios

or

B and C have opposite charges.

(ii) The particles are losing energy

or

speed

or

momentum is decreasing.

10. (a) $I = \dfrac{V}{R}$

$$I = \frac{12}{48}$$
$$I = 0.25 \text{ (A)}$$

(b)
$$E = -L\frac{dI}{dt}$$
$$-12 = -4.0\frac{dI}{dt}$$
$$\frac{dI}{dt} = 3.0 \text{ As}^{-1}$$

(c) X_L /(inductive reactance) increases

Or back emf **increases**

Therefore current decreases
(impedance increases)

11. (a) Coloured fringes are produced by **interference**.

    Reference to **different colours produced by**
    • angle of viewing
    or
    • thickness of film
    or
    • optical path difference

    (b) $d = \dfrac{\lambda}{4n}$

    $d = \dfrac{555 \times 10^{-9}}{4 \times 1.38}$

    $= 1.01 \times 10^{-7}$ m

    (c) Wavelengths in the middle of the visible spectrum not reflected or destructively interfere.

    Red and blue reflected / combined to (form purple).

12. (a) Division of wavefront

    (b) (i) $\Delta x = \dfrac{\lambda D}{d}$

    $\Delta x = \dfrac{510 \times 10^{-9}\ 2.5}{3.0 \times 10^{-4}}$

    $\Delta x = 4.3 \times 10^{-3}$ m

    (ii) % Uncertainty in $\lambda = \dfrac{2 \times 100}{510} = 0{\cdot}4\%$

    % Uncertainty in $D = \dfrac{0.05 \times 100}{2.5} = 2\%$

    % Uncertainty in $d = \dfrac{0.00001 \times 100}{0.0003} = 3.3\%$

    % Uncertainty in $\Delta x = \sqrt{2^2 + 3.3^2} = 3.9\%$

    Absolute uncertainty in
    $\Delta x = 3.9\% \times 4.3 \times 10^{-3}$
    $= 1.7 \times 10^{-4}$ m

    (iii) Slit separation.

    Highest **percentage** uncertainty.

13. (a) (i) The tablet emits plane polarised light.

    (ii) The brightness would gradually reduce from a maximum at 0 degrees to no intensity at 90 degrees.

    It would then gradually increase in intensity from 90 degrees to 180 where it would again be at a maximum.

    (b) $\tan \theta_1 = n$ $\theta_1 = Brewsters\ angle$

    $\tan \theta_1 = 1{\cdot}33$

    $\theta_1 = 53.1°$

    $\theta = 90 - 53.1 = 36.9°$